## *About the Book*

From fundamentals such as the model of the atom and chemical combinations of atoms to form molecules, to phenomena such as atom splitting and nuclear fission, science writer Melvin Berger gives a penetrating look at a timely and infinite subject. Expanding on the traditional picture of protons, neutrons and electrons, Berger probes deeply into the mystery of quarks, the tiny bits of matter discovered just a quarter century ago. He also examines recent scientific advances, from the manufacture of synthetic compounds to the development of high-tech accelerators for detecting infinitesimal particles.

Berger's thought-provoking outlook on the impact of physics on our future is complemented by sharp, accurate drawings illustrating the basics of atoms, as well as some at-home experiments where *you* can have the fun and adventure of being a modern-day physicist.

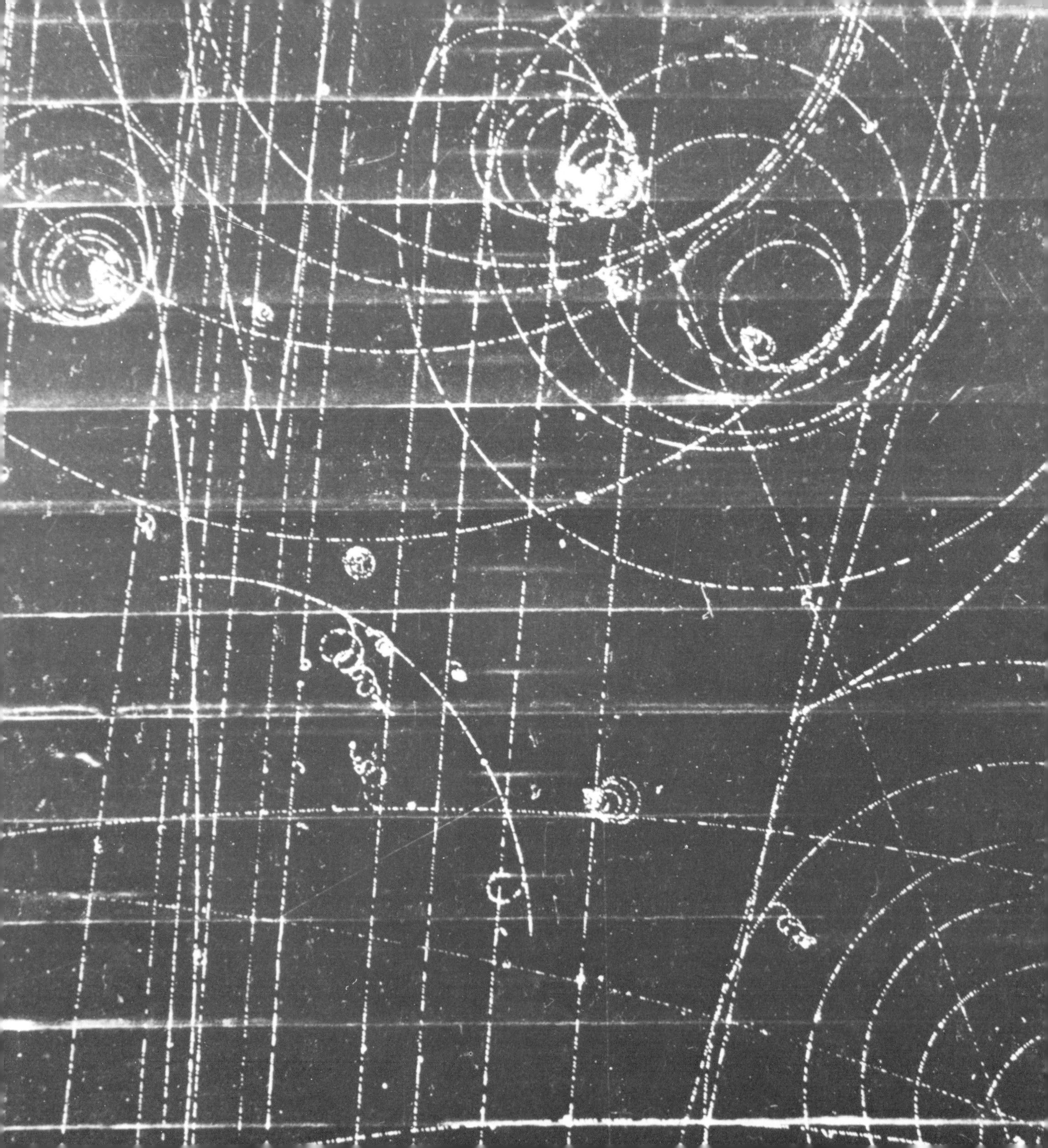

# ATOMS, MOLECULES and QUARKS

---

Melvin Berger

illustrations by Greg Wenzel

---

G.P. PUTNAM'S SONS / New York

*Photo Credits*
Cover photograph of a three-dimensional model of a polymer molecule was generated by a computer as part of an exploratory research project conducted by the General Electric Research and Development Center. The blue spheres represent the carbon backbone, the white spheres the hydrogen atoms and the red spheres mark the oxygen atoms. Courtesy, General Electric Research and Development Center, Schenectady, New York.

Photograph on page 2 of particle trails in a bubble chamber and photographs on pages 58, 59 and 60, courtesy Brookhaven National Laboratories.

 Published simultaneously in Canada by General Publishing Co. Limited, Toronto
Printed in the United States of America
Second Impression
Book design by Alice Lee Groton



Library of Congress Cataloging-in-Publication Data
Berger, Melvin.
Atoms, molecules and quarks.
Includes index.
Summary: An explanation of the composition, behavior, and uses of atoms, molecules and quarks, the building blocks of the universe.
1. Atoms—Juvenile literature. 2. Molecules—Juvenile literature. 3. Quarks—Juvenile literature. [1. Atoms. 2. Molecules. 3. Quarks]
I. Wenzel, Gregory, ill. II. Title.
QC173.16.B47 198 539.7 86-636
ISBN 0-399-61213-0

# CONTENTS

# INTRODUCTION

***What Is the World Made Of?***

People have always thought about the world around them. Through the centuries they have wondered about the huge variety of materials on earth—trees, stones, plants, metals, water and air, to mention just a few.

From this sense of wonder have come some questions: What are all these materials made of? In what ways are they the same? In what ways are they different?

Scientists have long sought answers to these questions and have made some startling discoveries. One of the first was that everything in the universe is made up of chemical *elements*. These are simple, natural substances that cannot easily be broken down into anything simpler. You are already familiar with several elements: the *gold* and *silver* in jewelry, the *oxygen* in the air, the *aluminum* in foil, the *neon* in signs, the *copper* in pennies and the *iron* in nails.

Researchers have also found that many materials are combinations of elements. Such combinations are called *compounds.* Compounds can be broken down into their separate elements. Sugar, salt, water, gasoline and nylon are just a few compounds you are most likely familiar with.

### *Atoms*

Every compound is made up of elements. And every element is made up of even more basic units called **atoms.** Each element contains only one kind of atom, so all the atoms of an element are basically the same.

Atoms are so tiny that they are invisible to the human eye. There are billions and billions of atoms in even the smallest speck of matter.

It is almost impossible to change an atom. Think of a ring made of the element gold. The ring contains only gold atoms, all of them identical. You could grind it into dust and each separate bit would still contain only complete gold atoms. You could freeze it, or heat it until it melted. But you would still have identical gold atoms and nothing else.

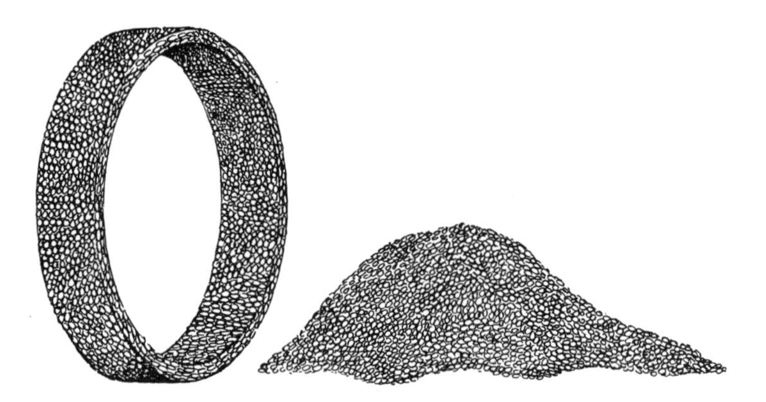

### *Molecules*

Most atoms, though, do not stay alone. They are joiners. They combine with other atoms to form **molecules.** A molecule is made up of two or more atoms.

Some *molecules* are composed only of atoms of one *element*. An oxygen molecule, for example, contains two oxygen atoms joined together.

An atom may also combine with one or more atoms of other, different elements. This forms *molecules* of a *compound*. The compound water consists of molecules containing two atoms of hydrogen joined with one atom of oxygen.

The results of combining different atoms to form a compound can be quite amazing. The compound can be completely different from the original atoms. For example, if you combine two poisonous elements, chlorine and sodium, you get the harmless compound table salt.

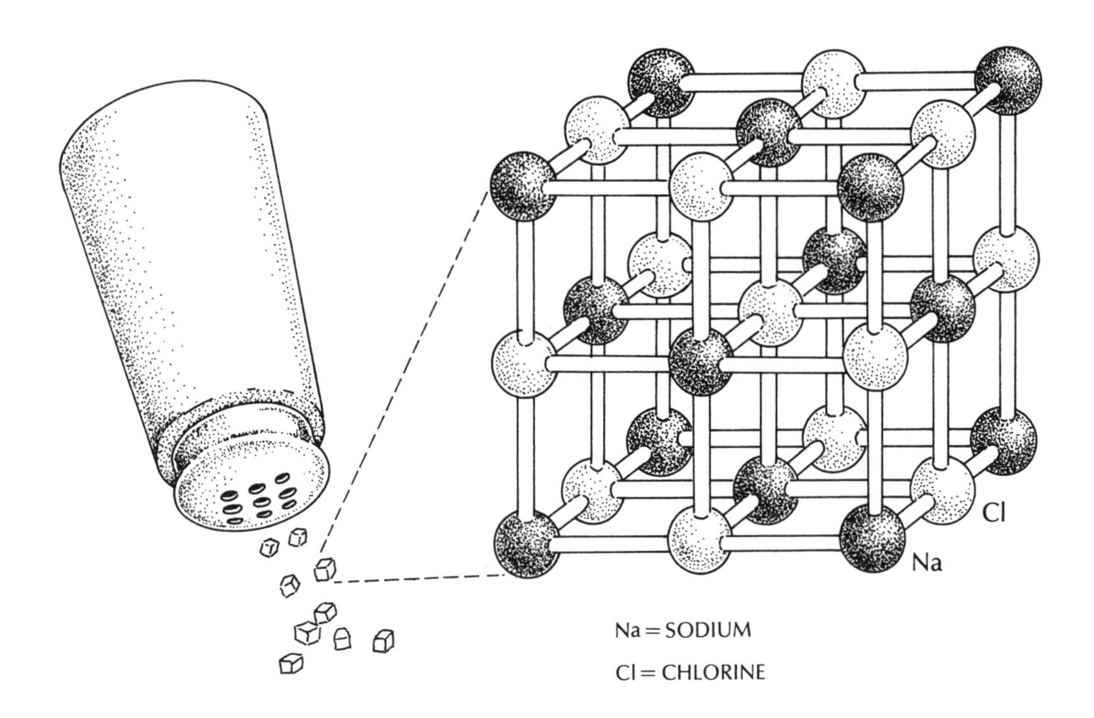

### *Quarks*

Atoms are so tiny that no one can see a single atom, even with the most powerful microscope. Yet, nearly one hundred years ago, scientists discovered that atoms are made up of even smaller pieces, called *particles*. Then, only about twenty-five years ago, they came up with the idea that within some of the particles are tiny, tiny bits, which they named **quarks.** Many scientists now believe that quarks may be the most basic units of matter.

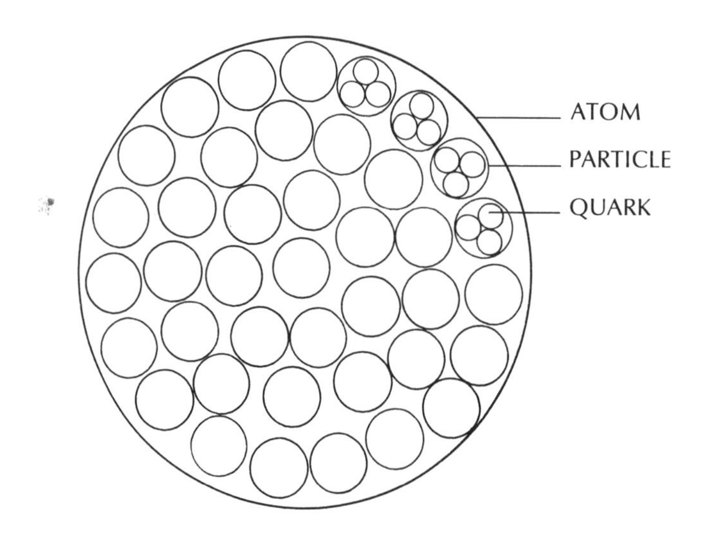


Scientists have traveled a long way, then, from the time people first asked the question: "What is the world made of?" The story of their discoveries is an exciting one, with many unusual twists and turnings. And the journey is not yet over. Far from it. Scientists are still plumbing the mysteries of those fascinating building blocks of the universe—atoms, molecules and quarks.

# 1. ATOMS

***What Atoms Are***

Atoms make up everything in our world. But it is hard to realize how small atoms really are. For example, do you know that if each person were as tiny as an atom, all the people alive today would fit on the head of a pin? Here is another way of thinking about atoms: There are as many atoms in a glass of water as there are grains of sand on all the beaches of the world!

It was the ancient Greeks who first had the idea that atoms are the smallest particles of matter. The philosopher Democritus (460–370 B.C.) coined the name *atom*. He took it from the Greek word *atomos,* which means "that which cannot be split." Along with other ancient philosophers he thought of the atom as a hard particle of matter that cannot be divided into smaller particles.

For more than two thousand years nothing happened to advance the Greek idea that matter is made up of tiny, indivisible atoms. Then, in the early 1800s, John Dalton (1766–1844), an English chemist, put forth a much broader theory. He made several points, most of which are still accepted today:

- All matter consists of small bits called atoms.
- All atoms of the same element (such as iron, gold or oxygen) are alike.
- Atoms of different elements are different.
- Atoms cannot be changed or destroyed. Chemical reactions always involve the entire atom.
- Atoms can be combined in units called molecules. Each molecule contains definite numbers of atoms of the same or different elements.

For nearly a hundred years, everyone accepted Dalton's views on the atom. Then, starting in the 1890s, an astounding series of discoveries drastically changed this simple picture of the atomic structure of matter. Scientists found that the atom *could* be divided into even smaller *particles*. And they learned that the atom is not a hard, solid particle as had been believed. In fact, atoms proved to be mostly empty space!

### *Inside the Atom*

Most of the matter, or *mass*, of the atom is concentrated in the small central part, or core. This core is called the *nucleus*. (The plural is nuclei.) Imagine a whole atom enlarged to the size of a domed football stadium with a golf ball hung right in the center. The golf ball in the middle of the immense stadium can be compared to the nucleus in the center of all the empty space of the atom.

Two basic kinds of particles, *protons* and *neutrons,* are found in the nucleus. The proton and neutron are about the same in size. The neutron, though, has slightly more mass. (Mass is related to weight. It is the resistance of an object to a change of motion.)

The main difference between the proton and the neutron is in their electrical charge. The proton has a positive electrical charge. This means that it is attracted by a negative charge and repelled by another positive charge. The neutron is neutral; it has no electrical charge at all.

The atom's nucleus, we said, takes up just a very, very small part of the space of the atom. But whirling around the nucleus, at very great distances from it, are much lighter particles called *electrons*. The orbiting electrons have a mass scarcely 1/2,000 that of the protons or neutrons. Electrons have a negative electrical charge equal to the positive charge of the protons.

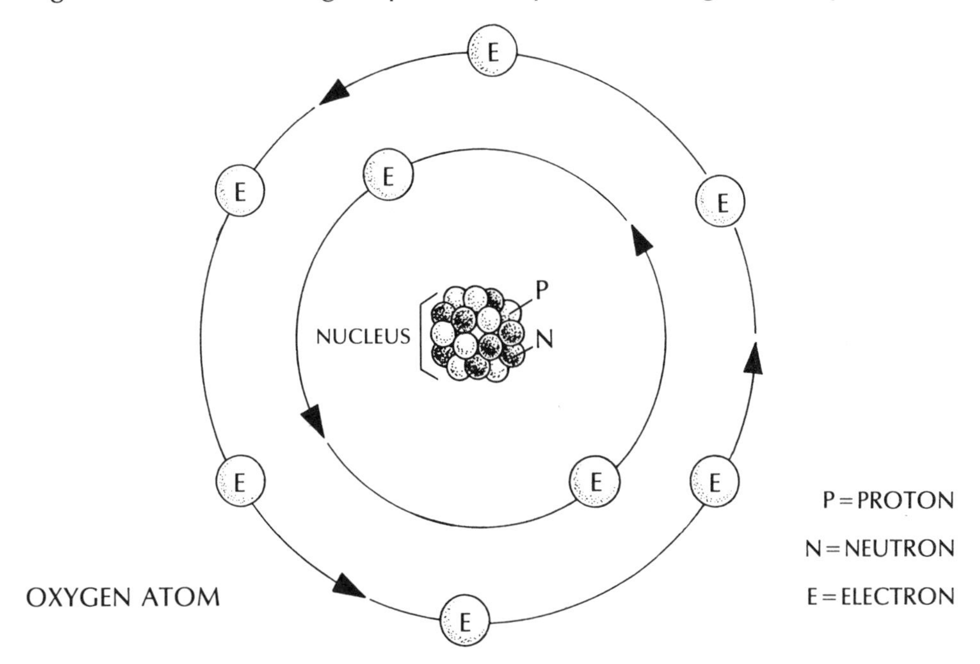


OXYGEN ATOM

In the atom, the number of electrons is ordinarily the same as the number of protons. The positive charge of the protons balances the negative charge of the electrons. This makes the atoms electrically neutral. The oxygen atom, for instance, has 8 protons and 8 circling electrons. In uranium, there are 92 protons and 92 circling electrons.

You can, though, throw the balance off in this simple experiment. You'll need a comb, some cloth made of nylon or Dacron (old stockings are fine) and a piece of newspaper or tissue paper torn into tiny bits. Rub the comb very briskly with the nylon. Do it for at least a half minute. The rubbing causes some of the electrons in the nylon to move onto the comb. Now hold the comb over the pieces of paper. What happens? The paper is drawn up to the comb.

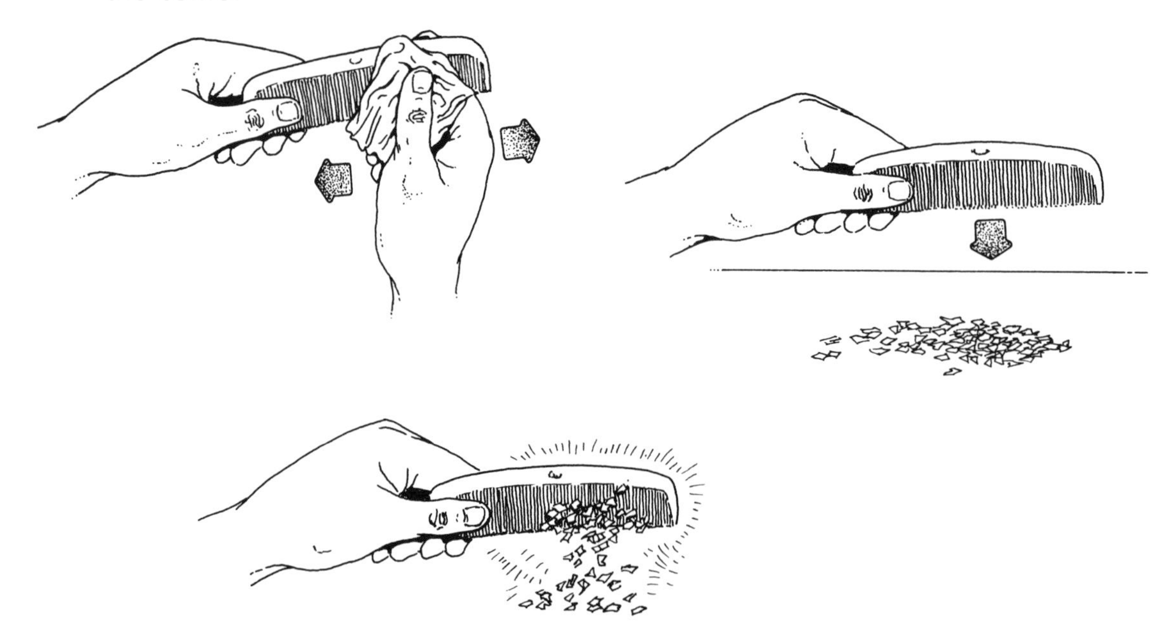

The extra electrons in the comb polarize some of the atoms in the paper, giving them a positive charge. As a result the paper, with its positive charge, is attracted to the comb, with its negative charge.

Since scientists cannot see, touch or weigh atoms, they surely cannot see, touch or weigh the particles within the atom. Yet they have been able to discover that atoms contain these three main particles—protons, electrons and neutrons.

How scientists found and proved the existence of the three basic particles is one of the most fascinating chapters in the history of science.

***Electrons***

The tool that gave science its first look within the atom was the Crookes tube, developed by the English scientist Sir William Crookes (1832–1919). The Crookes tube came in many shapes, but it was always made of glass. At each end there was a metal plate with an electric wire attached. It had a thin neck that was connected to a vacuum pump so that the air could be pumped out.

When the air was removed from the Crookes tube and the wires were connected to a strong source of electricity, there was a bright glow in the tube. The glow was called a cathode ray since it traveled from the negative metal plate (the cathode) to the positive plate (the anode).

Toward the close of the nineteenth century, Joseph John Thomson (1856–1940), director of the Cavendish laboratory at Cambridge University in England, set out to learn what he could about the cathode ray. In an early experiment, Thomson coated the anode in the Crookes tube with a fluorescent chemical that he knew would glow when struck by cathode rays. Then he put a metal cross in the path of the cathode rays. He saw a clear shadow of the cross on the anode. In this way, Thomson learned that cathode rays travel in straight lines.

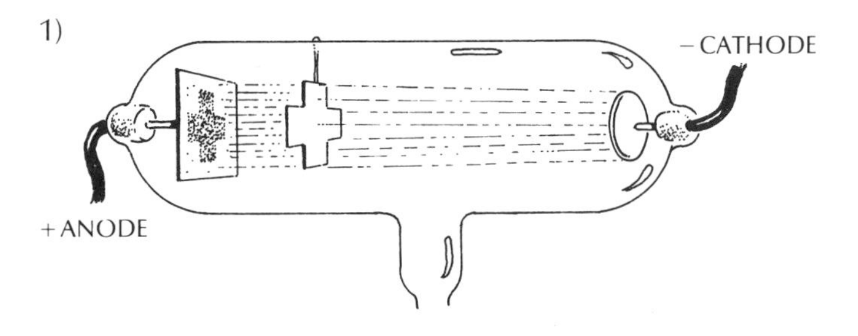

Next, Thomson put a delicately balanced paddle wheel in the path of the cathode rays. The cathode rays were able to spin the wheel. Thomson concluded that cathode rays are really made up of particles of matter, rather than beams of light.

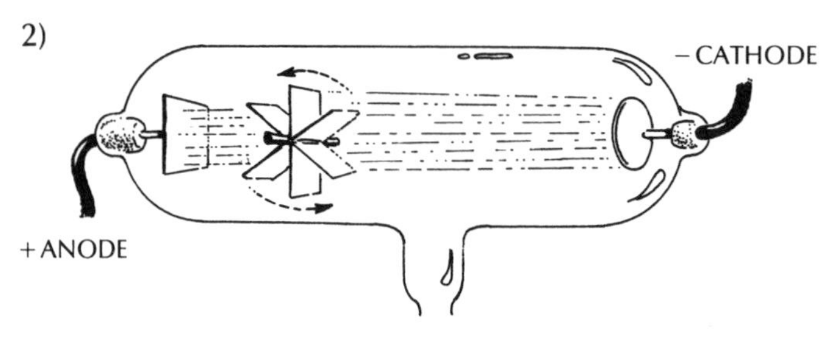


For the third experiment, Thomson placed the north and south poles of a magnet on either side of the Crookes tube. He noticed that the cathode rays were not attracted to either pole but were bent by magnetism. The direction the rays were deflected told him that they were a stream of particles with a negative electrical charge.

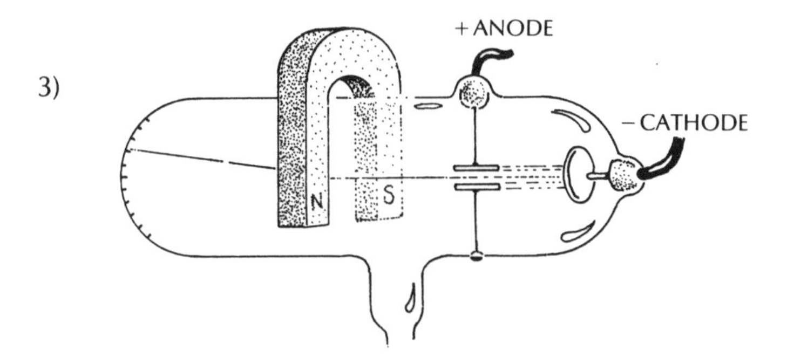


Fourth, Thomson put electrically charged plates on either side of the stream. By measuring the amount of charge necessary to bend the stream of particles, Thomson was able to calculate the mass of the particles. He found that the cathode particles were about 1/2,000 the mass of a hydrogen atom, the lightest known element.

In other experiments Thomson inserted different metal cathodes and traces of different gases in the tube. In each case, the particles behaved the same way. He guessed, therefore, that the particles were part of all matter, and were always the same.

4)

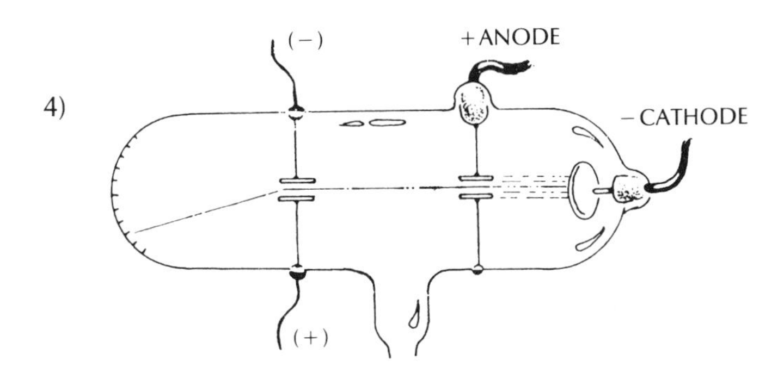


On April 30, 1897, Thomson announced, "Cathode rays are particles of negative electricity." Since these particles came from within the atom, he said, the "atoms are not indivisible, for negatively electrified particles can be torn from them by the action of electrical forces." And he concluded, "These particles are all of the same mass and carry the same charge of negative electricity from whatever kind of atom they may be derived, and are a constituent of all atoms." He called these particles electrons.

The atom, in Thomson's model, is something like a chocolate chip cookie, except that it is in the shape of a round ball. The entire mass of the cookie is the atom. It has a positive electrical charge. Scattered about are the electrons, the tiny bits of negative electricity. They are similar to the chocolate chips in the cookie. And the negative electrons exactly balance the positive mass of the atom, making the entire atom electrically neutral.

### *Protons*

Ernest Rutherford (1871–1937) was digging potatoes in a field in his native New Zealand when he got a letter from Thomson. The letter informed Rutherford that he had been accepted as Thomson's student. Rutherford threw down his spade and vowed never to dig another potato. What he didn't know was that he would be taking the next giant step toward unraveling the secret of the atom.

As Thomson's pupil, Rutherford wanted to find some answers to the questions of atomic structure. The best way to learn more about the inside of the atom, he decided, was to blow it apart.

Rutherford chose the nucleus of the helium atom as the bullet to shoot at the atom. The helium nucleus, which is called an *alpha particle*, contains two protons and two neutrons. The "gun" to fire the alpha particles was the element radium. Radium is radioactive. It is continually shooting out atomic particles, including the alpha particle. He placed the radium in a heavy lead container, with just a small opening to direct the escaping alpha particles.

The target for the alpha particles was a very thin sheet of gold foil, less than 1/100,000 of an inch thick. This is even thinner than the aluminum foil you use to cover food. Yet atoms are so small that the gold foil still had a thickness of more than 2,000 atoms!

For his first experiment, Rutherford set the gold foil in front of the radium container. Behind the foil he placed a fluorescent screen. The screen would show a spark of light whenever it was struck by an alpha particle. Thus, he could see whether any alpha particles were able to pass through the atoms in the gold foil.

Would you expect to be able to shoot a bullet (alpha particle) through a brick wall (gold foil) 2,000 bricks (atoms) thick? The actual results were amazing. Rutherford got flashes of light on the screen. Somehow the alpha particles were able to get through!

The scientist moved the screen to the sides and even in front, facing the

# RUTHERFORD'S GOLD FOIL EXPERIMENT

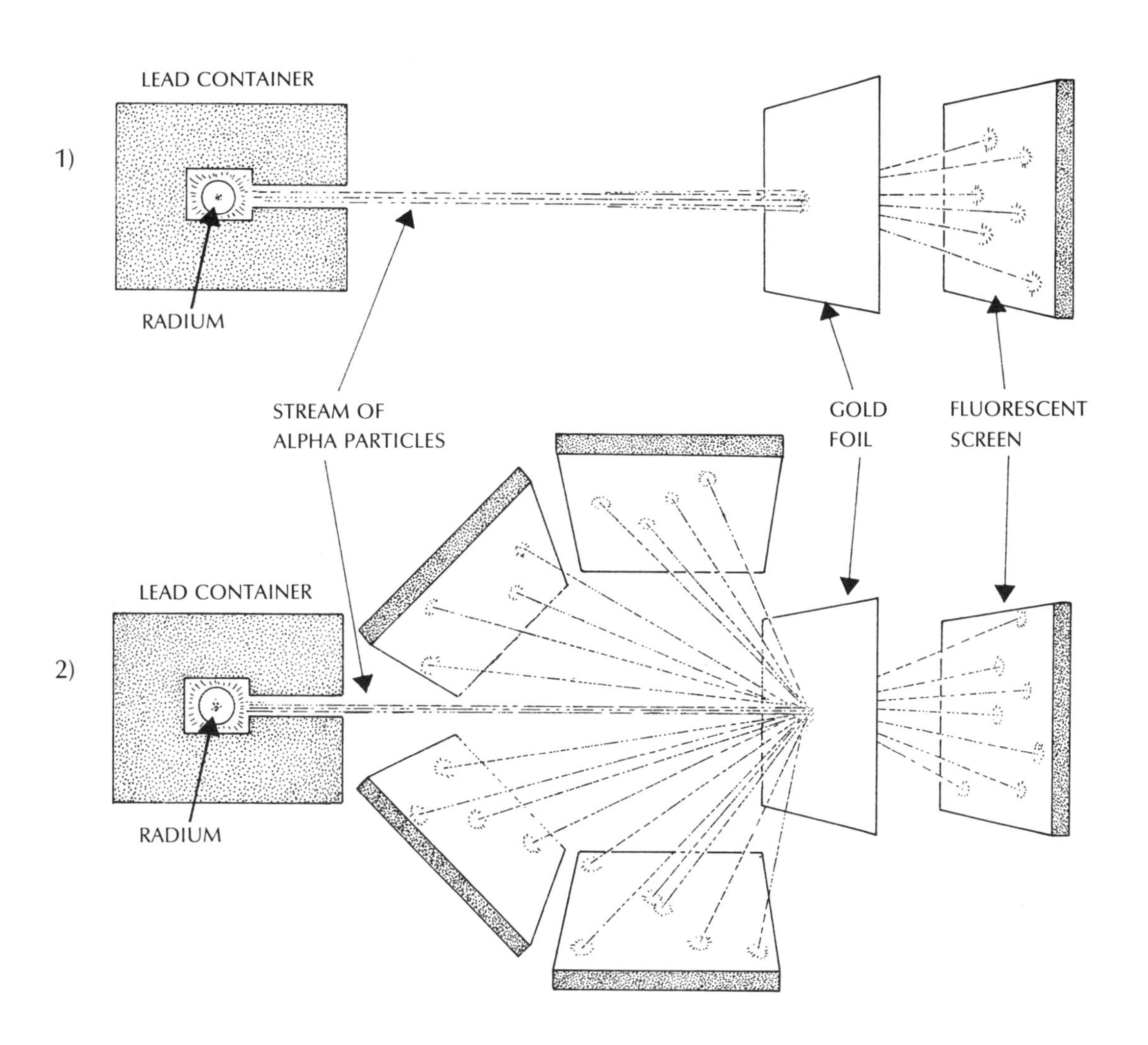

foil. To his amazement he found light flashes at all angles. Not only were alpha particles getting through, but some were bouncing off to the side as well as straight back at the radium. Rutherford wrote: "It was quite the most remarkable event that ever happened to me in my life. It was almost as incredible as if you had fired a 15-inch shell at a piece of tissue paper and it came back and hit you."

In 1911 Rutherford was finally able to explain what had happened. (The people working with him knew when he had solved a tough problem because at such times he sang "Onward Christian Soldiers" at the top of his lungs.) He suggested that the atom consists of a very small, heavy central core, called the nucleus. Very far out from the nucleus are the rapidly swirling electrons, each one traveling in a definite orbit.

Rutherford compared the structure of the atom to that of the solar system. He said that the nucleus is at the center of the atom, just as the sun is at the center of the solar system. At very great distances from the center are the orbiting electrons, just as the planets orbit at great distances from the sun.

The atom was largely an empty shell. That explained how the alpha particles were able to get through the gold foil. And within the shell there was a small though heavy nucleus, with a positive charge. That is what deflected some of the positive alpha particles and bounced back the few that actually hit the nucleus of the gold atoms.

Rutherford performed another experiment, similar to the first, to check his picture of the atom. But this time he used the element nitrogen instead of gold as the target of the alpha particles. As Rutherford expected, most of the alpha particles went straight through the empty space of the nitrogen atoms. A few particles bumped into the nucleus and bounced off.

At the end of the experiment, though, Rutherford detected the presence of hydrogen atoms. But these hydrogen atoms were different. They had a positive charge. (The hydrogen atom normally contains a single proton in the nucleus and one circling electron. If the electron is removed, all that is left is the proton, which gives the atom a positive electrical charge.) This led Rutherford to another very important idea. He realized that the hydrogen atoms had to have come from within the nitrogen atoms. He concluded, therefore, that the atoms of every element contain one or more of these positively charged hydrogen atoms. These positive hydrogen atoms are called *protons,* from the Greek word for first.

On the basis of these results, Rutherford set forth a more complete model of the atom. The nucleus, he said, is made up of heavy, positively charged protons. It has a positive electrical charge. Very far out from this nucleus are the much lighter electrons. Their negative charge balances the positive charge of the nucleus.

### *Neutrons*

Rutherford's solar system model was an important advance. But it could not account for the *motion* of the electrons within an atom. The electrons that circle the nucleus have negative electrical charges. And it was known that a rotating body with an electrical charge always gives off waves of energy. Within a tiny fraction of a second the energy of the electron is gone. It should fall into the nucleus. How can atoms exist if the electrons collapse into the nucleus?

One of Rutherford's students, the Danish physicist Niels Bohr (1885–1962), set about trying to solve this problem. He proposed that each electron had a definite energy level that had to do with its possible orbits. An electron could change its energy or orbit only in jumps, and only certain orbits were possible.

Bohr's explanation was based on the *quantum theory,* first stated by the German physicist Max Planck (1858–1947) in 1900. Simply put, the basic idea of the theory is that energy, which seems continuous, is made up of separate units called quanta (plural of quantum). This theory is related to the atomic theory, which holds that matter, which seems continuous, is made up of separate units called atoms. Another way of putting it is that energy, such as light or heat, does not come out in a continuous stream. It comes out in little packets or quanta.

Bohr was able to join Rutherford's model of the atom with Planck's quantum theory in 1913. He worked out three basic rules for the orbits of electrons around the nucleus: (1) The electrons can rotate around the nucleus in only a very few paths. (2) As long as they are traveling in one of these paths the electrons do not emit energy. (3) The electrons give off or gain energy (in the form of a single quantum) only when they jump from one path to another. Thus, as long as the electrons stay in their normal paths, the atom keeps its structure.

This was a more complete description of the atom. Yet the nucleus still remained a bit of a mystery. The nucleus of an atom had a greater mass than could be accounted for by protons. This problem led two scientists, each on his own, to come up with the same idea. Around 1920, Rutherford in England and William D. Harkins (1873–1951) in the United States both predicted that there was another particle in the nucleus. The scientists agreed that this other particle had the same approximate mass as the proton.

The two men also felt that the other particle had no electrical charge. A charge would upset the electrical balance between the negative electrons and the positive protons. Harkins described these particles as a "second, less abundant group with a zero net charge." He called them *neutrons* because they were electrically neutral.

Twelve years later, in 1932, Rutherford's pupil James Chadwick (1891–1974) proved the existence of the neutron. Chadwick was bombarding the element beryllium with alpha particles. Some particles, he noticed, were being knocked out of the beryllium. These particles were not turned by either pole of a magnet. He guessed they were electrically neutral. They were also able to knock protons out of other atoms. Therefore, they probably had a mass similar to that of the proton.

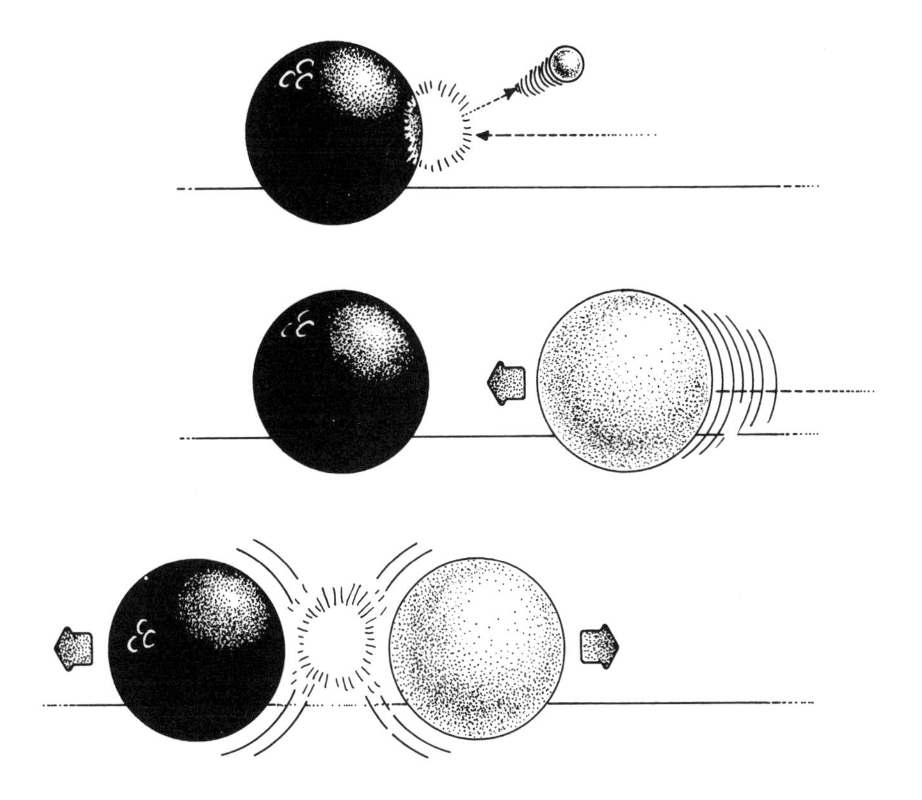

To understand Chadwick's reasoning, imagine throwing a Ping Pong ball at a bowling ball. The bowling ball does not move because the Ping Pong ball is so much lighter. But suppose you hit the bowling ball with another ball of about the same mass. This ball is able to move the bowling ball.

After Chadwick, the model of the atom was thought to be fairly complete. The nucleus of every atom, except the hydrogen atom, contains both protons *and* neutrons. The neutrons have about the same mass as the protons, but with no electrical charge. And there are electrons in orbit around the nucleus.

Does this mean that every atom is exactly the same? Not at all. All atoms of the same element have the same number of protons and electrons, but may vary in the number of neutrons. Atoms of different elements, though, have different numbers of each particle. In fact, it is the number of protons in the atoms that determines whether an element is lead or gold, hydrogen or oxygen.

### *Atoms and Elements*

The number of protons in the nucleus of an atom of an element is known as the atomic number of that element. Hydrogen is the lightest known element. It has an atomic number of 1 because it has only 1 proton in the nucleus. Carbon has 6 protons; therefore its atomic number is 6.

The atomic number is usually written to the left and below the letter symbol of an element. It tells you the number of protons in the nucleus. And it tells you the number of electrons in orbit around the nucleus, since the number of protons equals the number of electrons in an atom. Therefore, ${}_{1}H$ (hydrogen) contains 1 proton and 1 electron, ${}_{6}C$ (carbon) has 6 protons and 6 electrons, and ${}_{92}U$ (uranium) has 92 protons and 92 electrons.

The atomic mass of an atom is equal to the mass of its nucleus. In other words, it is the sum of the masses of its protons and neutrons. (Electrons have so little mass that they are not added in.) The mass number is usually written above and to the left of the symbol of the element. By adding the protons and neutrons, you can calculate the mass of the element.

Hydrogen (${}^{1}_{1}H$), we said, is the lightest element. It has a mass number of 1 for its single proton in the nucleus. The symbol for oxygen is ${}^{16}_{8}O$. The symbol tells you there are 8 protons and 8 neutrons in the nucleus and 8 electrons in orbit. The symbol for uranium 235 is ${}^{235}_{92}U$. It contains 92 protons, 143 neutrons and 92 electrons.

The atomic masses of all atoms of the same element, however, are not the same. They all have the same number of protons and electrons, so that they have the same chemical properties. But they contain different numbers of neutrons, which gives them different atomic masses. Atoms that are chemically alike but have different atomic masses are called *isotopes*. All elements have at least one isotope that occurs naturally. Many elements have more.

Uranium, you remember, has the mass number 235 (92 protons, 92 electrons, 143 neutrons). But an isotope of uranium, uranium 238, has 92 protons, 92 electrons and 146 neutrons. Chemically it is similar to uranium 235, but it is an isotope because of the three extra neutrons.

With the discovery of the neutron in 1932, the essential picture of the atom was complete. (Many more particles have been found since then. They are discussed in Chapter Three.) Still scientists continued their experiments on the atom.

### *Splitting Atoms*

One very important line of research was begun by Rutherford in 1920. Rutherford found that nitrogen could be changed into oxygen by bombarding nitrogen atoms with alpha particles. (Alpha particles, you will remember, are the nuclei of helium atoms; they contain two protons and two neutrons.) The nucleus of the nitrogen atom contains 7 protons and 7 neutrons ($^{14}_{7}N$). When the nitrogen atom is struck by the alpha particle, one proton and two neutrons from the alpha particle enter the nucleus, making a total of 8 protons and 9 neutrons ($^{17}_{8}O$), which is an isotope of oxygen. The nitrogen atom actually becomes an atom of oxygen!

Rutherford had fulfilled an ancient dream. He had changed one element into another. Later scientists learned how to do the same to many other elements. And then, in more recent times, they learned how to change atoms by splitting apart their nuclei. Another name for splitting atomic nuclei is nuclear fission.

In nuclear fission the two fragments that result have less mass than the original nucleus. What happened to the missing mass? Scientists found a clue in Einstein's special theory of relativity. Albert Einstein (1879–1955) showed that a small amount of mass can become a great deal of energy. His famous equation, $E = mc^2$, tells that the energy (E) of any particle of matter equals its mass (m) times the speed of light (186,000 miles per second) multiplied by itself ($c^2$). When an atom is split in nuclear fission the lost mass is changed into an explosive burst of heat, light and other high-energy radiation.

One of the first elements to have its atoms split was uranium. Because uranium has a large, heavy nucleus, it can release a tremendous burst of energy.

Of the uranium found in nature, 99.7 percent is uranium 238 (92 protons, 146 neutrons). When an atom of uranium 238 is struck by a neutron, it captures the neutron, and becomes uranium 239 (92 protons, 147 neutrons). A tiny fraction of the natural uranium, though, is uranium 235 (92 protons, 143 neutrons). When uranium 235 is struck by neutrons, the nucleus of the uranium is actually split into some two hundred fragments. The fragments include large numbers of atoms of barium (56 protons, 81 neutrons) and krypton (36 protons, 48 neutrons), and released neutrons.

The combined mass of the barium and krypton nuclei, plus the neutrons, however, is slightly less than the original mass of the uranium nucleus. That lost mass, according to Einstein's equation, is changed to energy. That is why there is an immense release of energy every time a uranium atom is split.

But the fission of uranium also releases more neutrons. These neutrons, in turn, split more uranium atoms, which release still more energy and neutrons. A reaction like this, which continues on its own, is called a chain reaction. The only way to stop it is to introduce some material that absorbs neutrons, such as cadmium or boron, into the flow of neutrons. This control material captures the free neutrons, preventing them from striking and splitting any other uranium atoms.

When uranium 235 atoms are hit with neutrons, only a small number of atoms are changed to energy. Not every nucleus of every atom is split. Great quantities of energy are released, however, by those few that are split.

This activity can show you how nuclear fission and chain reactions work. You'll need a large sheet of poster paper, a piece of cardboard, a colored marker and scissors. First cut out three cardboard disks about 3 inches across. These disks will represent uranium 235 atoms. Mark them $^{235}U$. Cut out two more disks about 2 inches across. These disks will be the barium and krypton. Mark one Ba and one Kr. Finally cut out three 1-inch disks. They will represent neutrons. Mark them n.

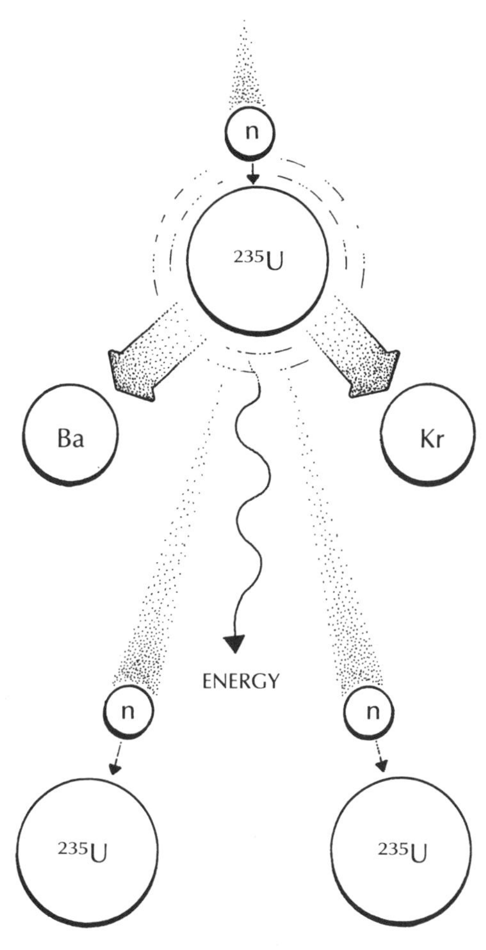

Paste a $^{235}U$ disk and an n disk near the center of the top of the paper, with the n above the $^{235}U$. Now draw an arrow showing the n about to strike the $^{235}U$ atom.

When a neutron strikes a uranium 235 atom, the atom splits. As a result, 1 barium atom and 1 krypton atom are given off. Paste the Ba and Kr disks off to the sides, with drawn arrows showing that they came from the $^{235}U$.

There is also a release of energy. Show the energy with a wavy arrow coming down from the $^{235}U$. And there are neutrons emitted that strike other uranium atoms. Paste the other two $^{235}U$ disks in the lower corners, and draw arrows from the first $^{235}U$ disk to the two others. Paste the other two n disks at the tips of these arrows.

The energy released from nuclear fission can be used in two ways. In a nuclear reactor, the chain reaction is controlled so that the splitting atoms just release great amounts of heat. This heat is then used for purposes like generating electrical power, just as the heat of burning oil, coal or gas is used to produce electricity. If the chain reaction is allowed to go on without control the result is a powerful explosion. This is the basic principle of the atom bomb.

### *Combining Atoms*

In the early 1950s scientists unleashed an even more powerful source of energy from within the atomic nucleus. This new source works by combining small atoms, not splitting large atoms. It is called nuclear fusion. Nuclear fusion on an immense scale is going on within the sun all the time. It is the source of all the heat and light energy pouring out of the sun, and all the other stars as well. And the hydrogen bomb uses nuclear fusion to produce a super-powerful explosion.

Nuclear fusion uses the lightest of all elements, hydrogen. But it is not the familiar hydrogen with an atomic mass of 1. Fusion works better with two heavier isotopes of hydrogen, deuterium and tritium. Fusing the nuclei of deuterium (one proton, one neutron) and tritium (one proton, two neutrons) results in helium (two protons, two neutrons; also called an alpha particle). The extra neutron and a gigantic burst of energy are released when the two nuclei fuse.

Scientists have run into a number of problems in their efforts to develop nuclear fusion as a practical energy source. The basic difficulty is that the only way to fuse the two hydrogen nuclei is to bring them to a temperature measured in millions of degrees. (Because of the extremely high temperature required, fusion is also known as a thermonuclear reaction.) The high temperature is necessary to overcome the repulsion between the two positively charged hydrogen nuclei that must combine for fusion to take place. Then they must find a way to contain the super-hot hydrogen gas (called a plasma at this heat). Researchers are currently experimenting with powerful magnetic fields and with short bursts of very strong laser beams to heat the hydrogen and to contain the fusion reaction.

You can show the fusion of hydrogen atoms into helium atoms using a large sheet of poster paper, cardboard, a colored marker and scissors. Cut out of the cardboard one 3-inch disk, two 2-inch disks and one 1-inch disk. The large disk is an atom of helium. Show the atom with 2 protons and 2 neutrons tightly packed in the nucleus in the center, and 2 circling electrons. Label the disk $^{4}_{2}He$. The two smaller disks are deuterium (1 proton and 1 neutron in the nucleus, 1 circling electron; $^{2}_{1}H$) and tritium (1 proton and 2 neutrons in the nucleus, 1 circling electron; $^{3}_{1}H$). Mark the smallest disk n for neutron.

Paste the two hydrogen disks near the top corners of the paper. Draw arrows from each one, meeting in the middle. From the point where the arrows meet, draw an arrow going off to one side. Paste the n at the tip of that arrow. The n is the neutron that is produced during fusion. Draw a wavy arrow heading off to the other side. Label it energy. It shows the release of energy. And finally make a very short arrow pointing straight down. Paste the helium disk there. That represents the fused helium atom.

There will never be a shortage of fuel for nuclear fusion. Deuterium and tritium can be drawn from sea water. That is why scientists are continuing their research. Eventually they hope to perfect nuclear fusion. It is one of the most promising energy sources of the future, and despite all the difficulties, many scientists believe that we will be using nuclear fusion as a major source of power before the end of the century.

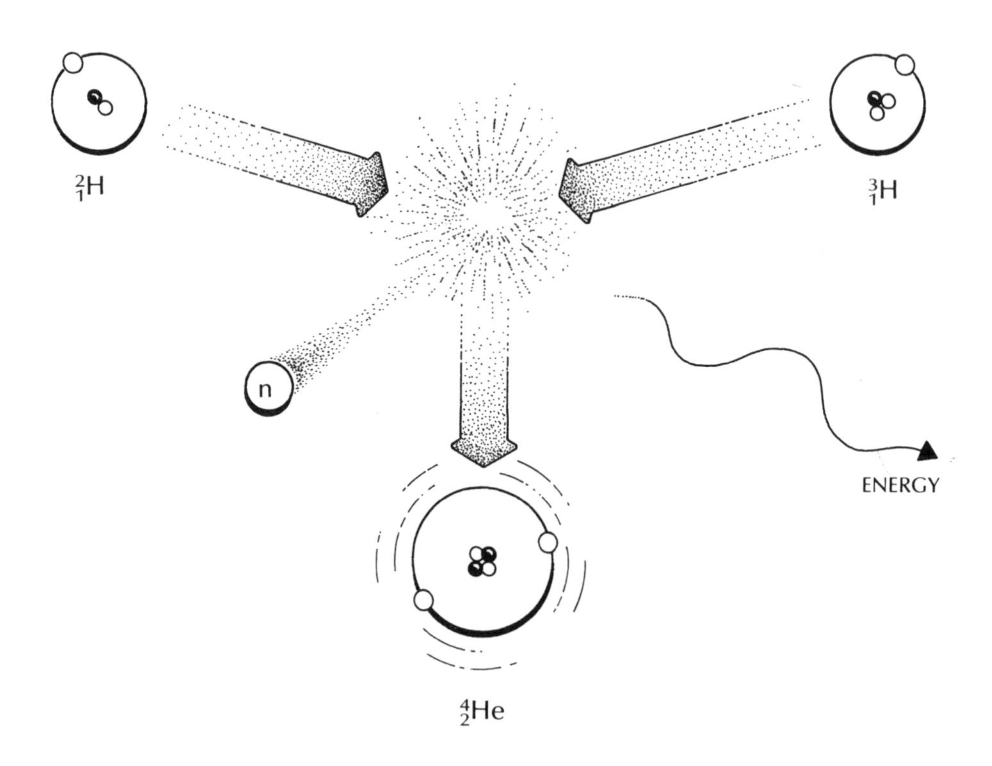

# 2. MOLECULES

***What Molecules Are***

Molecules are chemical combinations of atoms. That is, they are made up of two or more atoms. Atoms of the same element combine to form molecules of that element. Atoms of different elements combine to form molecules of compounds.

There are only about one hundred different kinds of atoms. But there are thousands of different kinds of molecules. Think of atoms as the letters of the alphabet and molecules as the words that can be made from the letters. And just as words are made up of different combinations of letters, molecules are made up of different combinations of atoms. Thus, salt molecules are different from sugar molecules because they contain a different combination of atoms.

Molecules are so small that there are billions of them in a single tiny grain of salt. Scientists can see only the very largest molecules under the most powerful microscopes. Yet they know that molecules are there.

To understand molecules, we should get to know what they are made of, what forces hold them together, how they behave and the different shapes they can take.

### *What Molecules Are Made Of*

Molecules of elements are made up of two or more of the same kind of atoms. Each molecule of oxygen, for instance, is made up of two atoms of oxygen. Hydrogen and nitrogen are examples of other well-known elements that contain two atoms of the same element in each molecule.

The molecule of a compound is made up of two or more *different* kinds of atoms. There are thousands of different compounds. Many, such as water, sugar and salt, occur in nature. Others, such as nylon, Dacron and the plastics, are manmade.

The formula for a compound shows what kinds of atoms it contains and also how many of each kind. The scientific name for salt, for example, is sodium chloride. Each molecule of the compound salt contains 1 atom of sodium (Na) and 1 atom of chlorine (Cl). It therefore has the formula NaCl. Water has the formula $H_2O$, which shows that 2 atoms of hydrogen (H) and 1 atom of oxygen (O) are joined to make up every water molecule. Sulfuric acid ($H_2SO_4$) has a molecule containing 2 atoms of hydrogen, 1 atom of sulfur and 4 atoms of oxygen. Sugar is a giant compound with many atoms. Each sugar molecule contains 12 carbon atoms, 22 hydrogen atoms and 11 oxygen atoms. Its chemical formula is $C_{12}H_{22}O_{11}$.

A compound can be broken down into its elements by chemically separating the different atoms. Even though you cannot handle a single molecule of sugar, you can separate a huge number of sugar molecules into atoms of carbon, hydrogen and oxygen. Breaking down the sugar will help you prove that various kinds of atoms came together to form the substance.

Place a tablespoon of sugar in the center of an old pie tin and set the pie tin on the kitchen stove. Place a small, clean glass jar upside down over the sugar. Warm the sugar slowly over a low heat.

As the sugar begins to melt you will see a cloud of steam form in the jar. Drops of liquid will appear on the sides. After a while, use a potholder to

remove the jar. Set it right side up. The liquid on the sides will run down to the bottom. Dip your finger into the liquid and taste it. What does it taste like?

It is water. Water, you know, is a compound made up of atoms of hydrogen and oxygen joined into molecules of water. Therefore, heating the sugar separated out the hydrogen and oxygen, not as individual elements, but combined as water molecules. (In a chemistry laboratory you would be able to separate the water into hydrogen and oxygen.)

Continue heating the sugar, watching it carefully. You will see it turn dark brown and then black. When it is completely dry, turn off the heat. Give the substance in the pie tin a few minutes to cool off. Now touch it with your finger and taste it. Does it taste sweet like sugar? No, it does not. The black substance you are tasting is carbon. It is the other element found in sugar, along with hydrogen and oxygen.

So you see, the characteristics of a compound can be quite different from the elements it contains. Sugar is a white, sweet-tasting solid. But it consists of hydrogen and oxygen, two gases that combine to make liquid water, and carbon, a black solid.

A compound can also be built up from atoms. The rust that forms on iron is a compound called iron oxide. It is made up of two elements—iron (Fe) and oxygen (O). It has the formula $Fe_2O_3$, indicating that each molecule contains 2 atoms of iron and 3 of oxygen. You can make iron oxide using steel wool for the iron and air for the oxygen, since about one-fifth of the air is oxygen.

Take a large glass bottle and drop in a pad of soapless steel wool. The bottle already has air in it, of course. Add a little water and shake the bottle to moisten the steel wool. The water will help to speed up the experiment. Cap the bottle and set it aside.

Look at the steel wool every day. In a few days you'll see some changes in the steel wool. The shiny gray metal begins to take on a red-brown coating. That coating is rust. The atoms of iron are combining with the oxygen in the air to form a new substance, rust, or iron oxide.

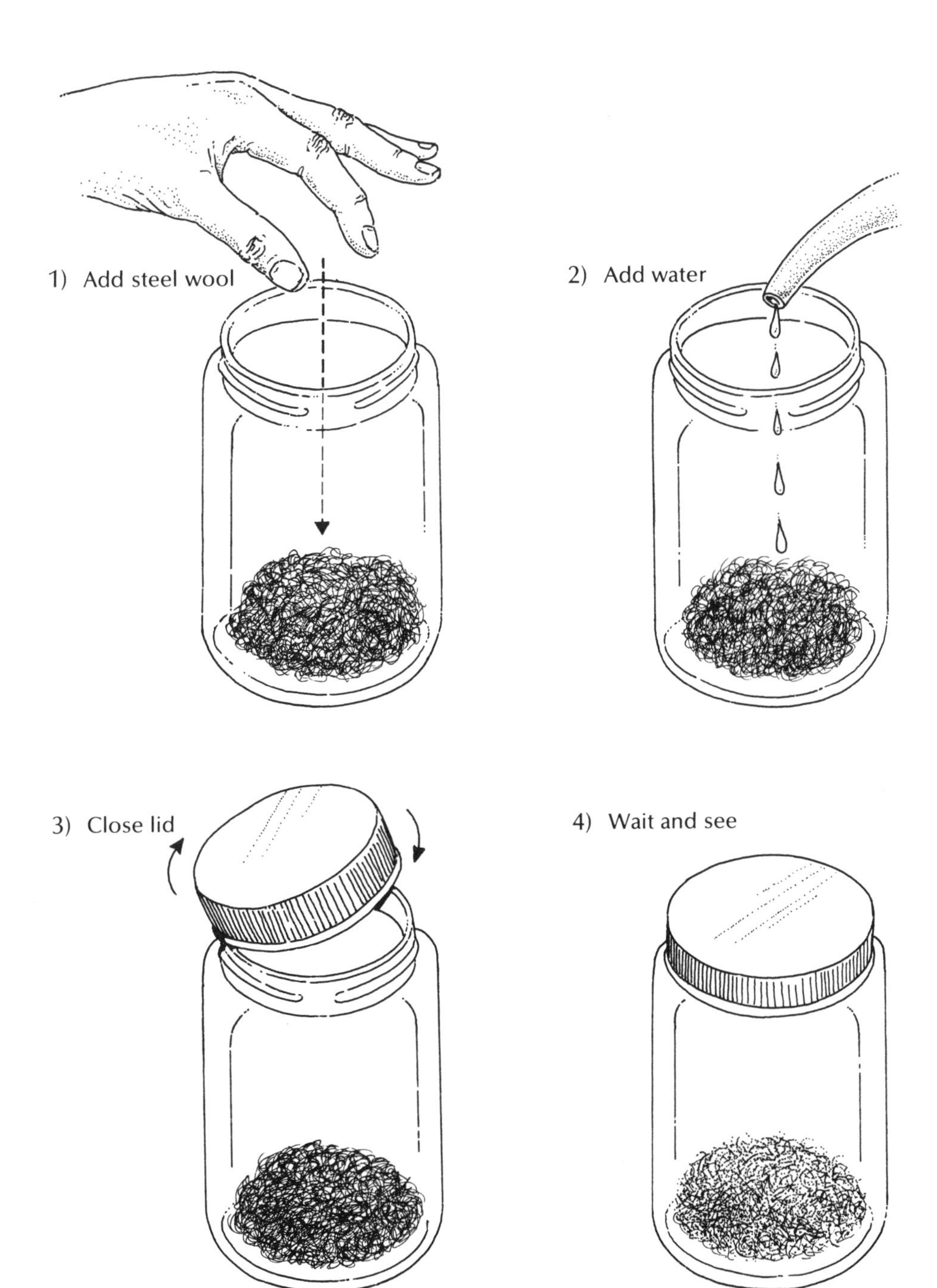
1) Add steel wool
2) Add water
3) Close lid
4) Wait and see

***What Holds Molecules Together***

Molecules are held together by a force, or energy, that comes from the electrons in the atom. The electrons, you recall, circle the nucleus in orbits. How or whether an element combines with other elements depends on the arrangement of the electrons in their orbits. The force that holds them together is called electron bonding.

Electron bonding can work in two ways: electrovalent or covalent. In *electrovalent* bonding there is a transfer of an electron from one atom to another. In *covalent* bonding the atoms actually share electrons. The particular type of bonding that occurs is usually the one that produces the greatest stability for the atoms involved.

Salt is a good example of electrovalent bonding. As you know, salt contains the elements sodium and chlorine. Sodium has 11 electrons in the three shells that surround its nucleus: 2 electrons are in the shell closest to the nucleus, 8 in the next shell and 1 in the shell farthest from the nucleus. Chlorine has 17 electrons in its three shells: 2 are in the first shell, 8 in the second and 7 in the outside shell.

The study of the motion of electrons showed scientists that atoms are most stable when they have 2 electrons in the first shell and 8 electrons each in the second and third shells. Therefore, the sodium atom has an extra electron in its third shell. And the chlorine atom is missing an electron from its third shell.

Can you see how simple it is for the sodium and chlorine atoms to become stable by combining? The single electron in sodium's outer shell fills the gap in the outer shell of the chlorine atom. This gives the sodium atom two complete shells and the chlorine three complete shells. Both atoms are now stable. And the result is a molecule of the compound salt.

But what holds the two separate atoms together in the salt molecule? As you know, sodium, like all other atoms, is electrically neutral. Its 11 positive protons exactly balance its 11 negative electrons. But the situation changes

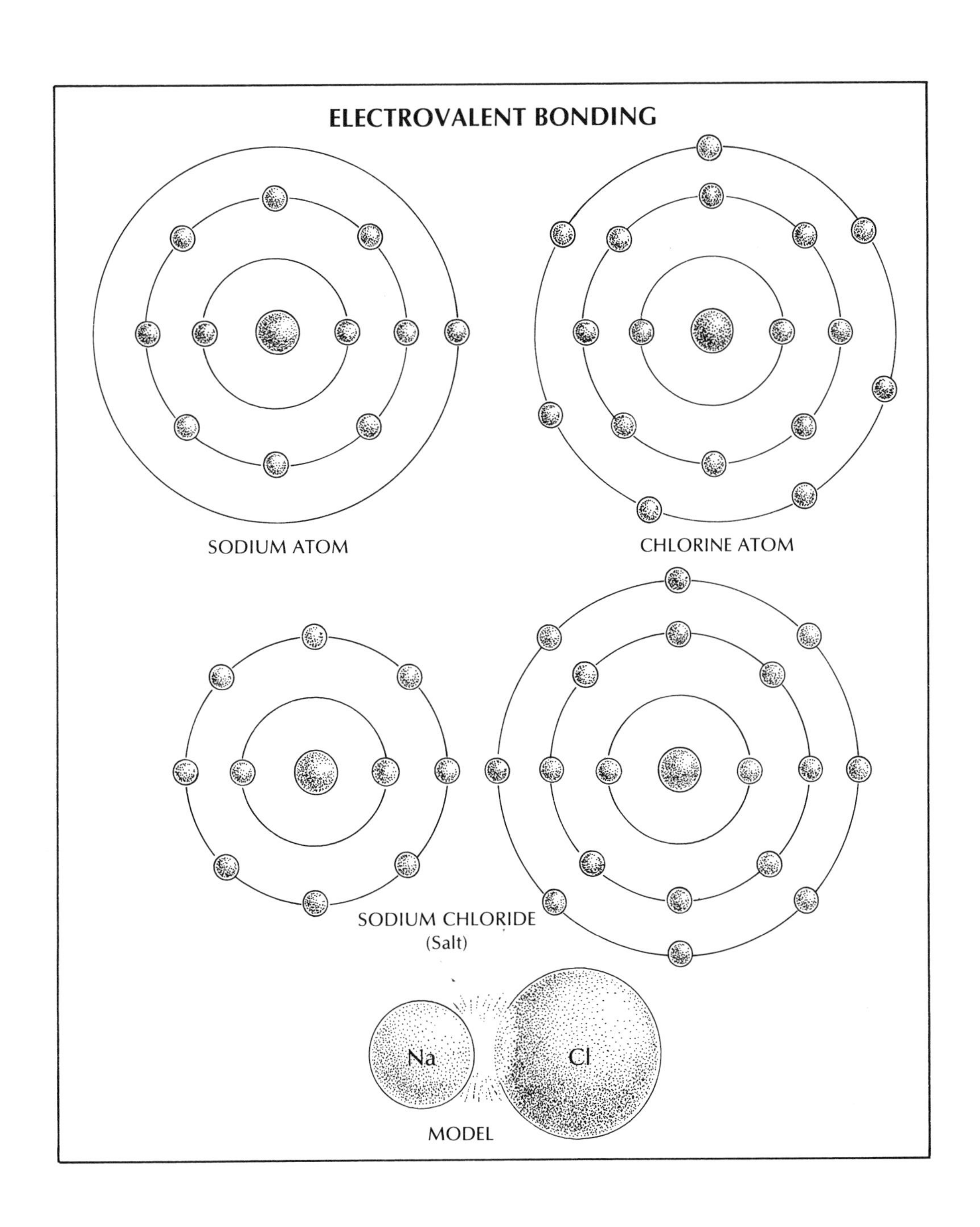
ELECTROVALENT BONDING
SODIUM ATOM
CHLORINE ATOM
SODIUM CHLORIDE
(Salt)
Na
Cl
MODEL

when the sodium atom loses an electron to the chlorine atom. The sodium now has more protons than electrons. It is left with a positive electrical charge. An atom that has an electrical charge, whether positive or negative, is called an *ion*. Thus, the sodium atom becomes a positive ion.

The chlorine atom also starts out electrically neutral, with 17 protons and 17 electrons. But it gains 1 electron from the sodium. The electrons now outnumber the protons, giving the atom a negative charge. Therefore, the chlorine is a negative ion.

You already know that opposite electrical charges attract each other. Thus the positive sodium ion and the negative chlorine ion are pulled together. This attraction is the electrovalent bond that keeps the salt molecule intact. This is also called *ionic bonding*.

More common than electrovalent bonding is covalent bonding. Water is a good example of this second way of forming molecules. You already know that each molecule of water contains 2 hydrogen atoms joined to 1 oxygen atom. Each hydrogen atom has only 1 electron. Oxygen has 8 electrons—2 in the first shell and 6 in the second shell. For the atom to be most stable, you recall, there need to be 2 electrons in the first shell and 8 in the second.

The oxygen atom will become stable if it gets 2 more electrons for its second shell. It gets these 2 electrons from the 2 hydrogen atoms that join the oxygen atom and share their electrons with it. The 2 electrons are still part of the hydrogen atoms. But they are also part of the oxygen atom's outer shell. That gives the oxygen atom a complete second shell with 8 electrons. With shared electrons, the oxygen and hydrogen atoms are now joined in the stable water molecule.

The force holding the atoms together in the water molecule is known as covalent bonding. A great many molecules, such as sugar, are formed in this way. In covalent bonding, the electrons never really leave their original atoms. The force that attracts the electrons to the nucleus also holds the atoms together in the molecule.

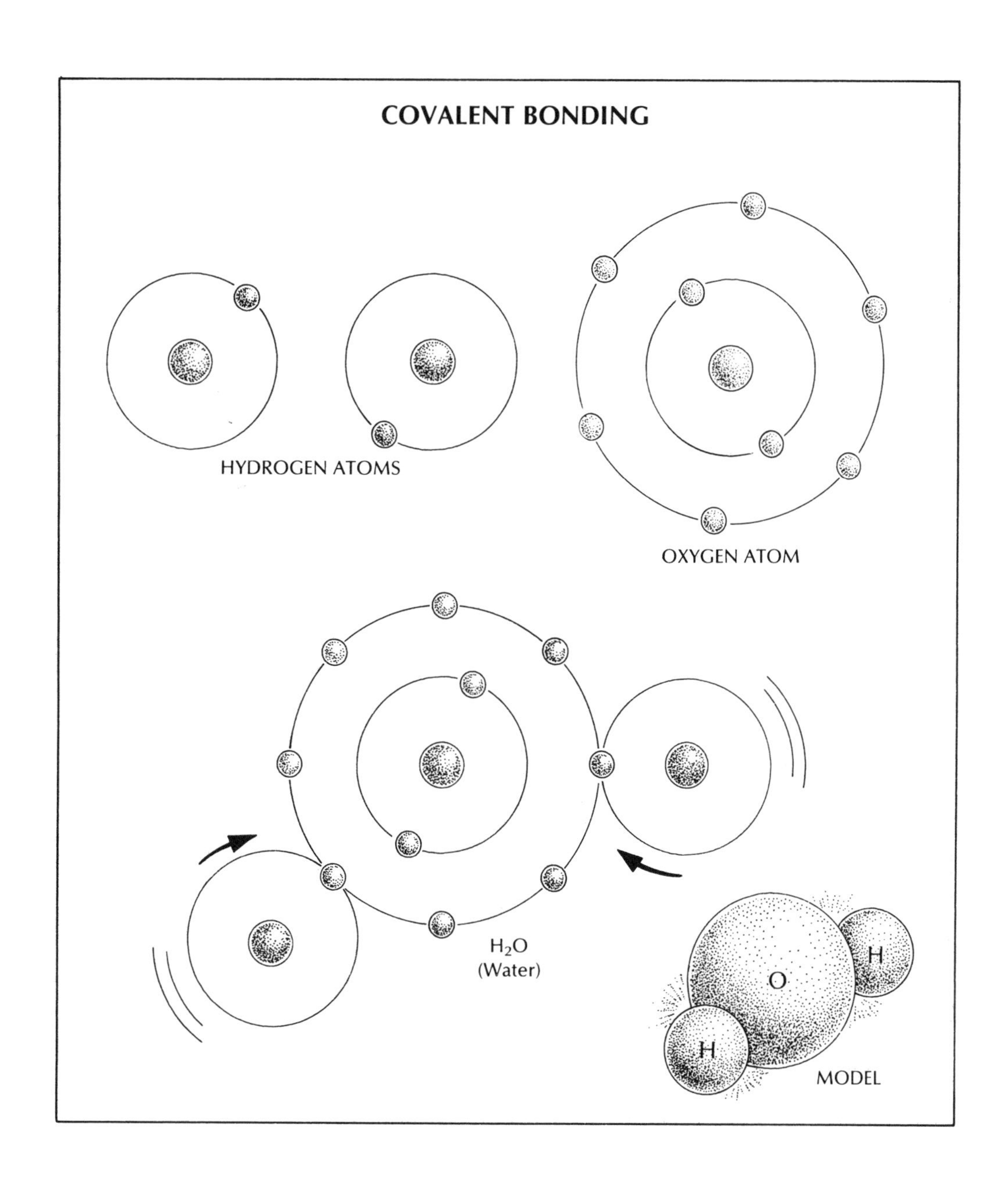
COVALENT BONDING
HYDROGEN ATOMS
OXYGEN ATOM
$H_2O$
(Water)
O
H
H
MODEL

### *Other Molecular Forces*

Molecules are also held together by two outside forces known as cohesion and adhesion. *Cohesion* is a force that attracts molecules of the same kind to one another. *Adhesion* is a force that causes the molecules of two unlike substances to stick together when they come close to one another.

Cohesion differs in different types of materials. It is strongest between molecules of solid substances, such as brick or iron, and weaker between molecules of liquids. In gases there is scarcely any cohesion at all.

In some materials the force of cohesion is so strong that a large force may be needed to pull the molecules apart. Pieces of steel or stone, for example, are all held very tightly together. You cannot tear apart even a thin piece of steel or push your hand into the softest stone. Each needs a good deal of force to overcome the cohesion and break it apart.

In other materials, cohesion is so slight that only a small force is needed to pull the molecules apart. It is very easy, for example, to tear apart the molecules in a piece of paper or to push apart the water molecules in a swimming pool. And you are hardly aware that you are disturbing the molecules as you move through the air.

**COHESION**

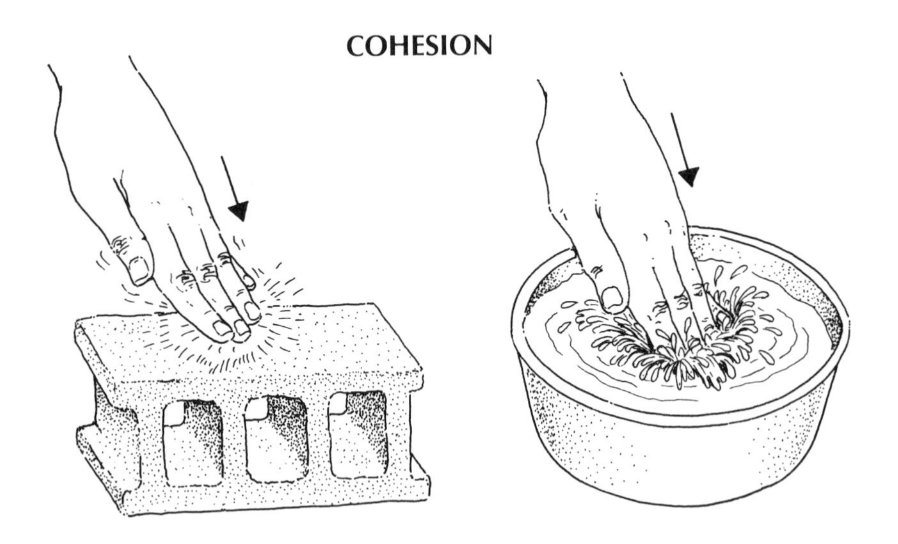

# ADHESION

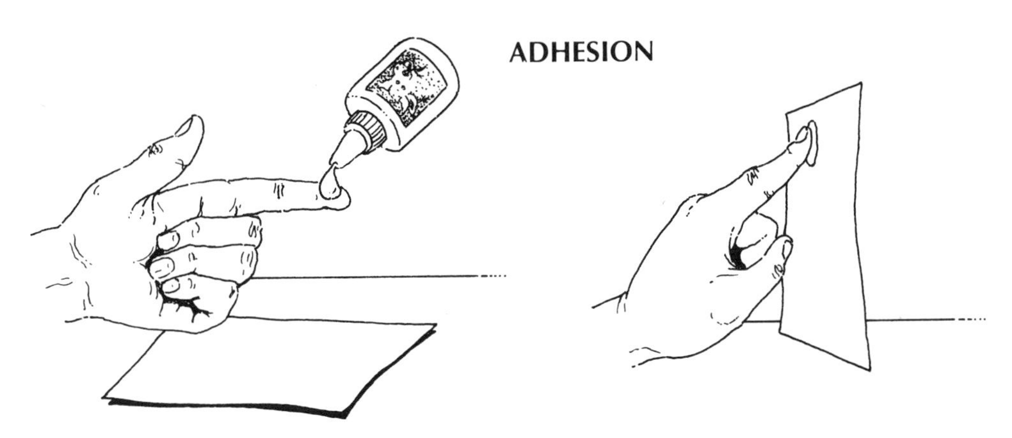

Adhesion, the force that attracts molecules of unlike substances, is a common property of matter. It causes chalk to stick to the blackboard, ink to stick to paper and dirt to stick to your hands. But the force of adhesion varies among different substances. Objects made of wood, paper, cloth or metal, for instance, have weak adhesion. They will not usually stick to one another. But a substance like paint will stick to all of them. Other materials, such as glue or paste, have strong adhesive properties. The force of adhesion between molecules of glue and molecules of most other substances is strong enough to make them stick tightly together.

Without the molecular forces of cohesion and adhesion, materials would not stick together. There would be no solids or liquids in our world. All molecules would fly apart and drift away like air. Every substance would look and feel like a gas.

### *How Molecules Behave*

All matter consists of a large number of molecules that stick together, but with spaces between them. These are literally empty spaces containing nothing, so that there is a vacuum between the molecules. Molecules in solids, like pieces of wood or metal, are packed quite closely together. They do not move much. Their shape does not change easily.

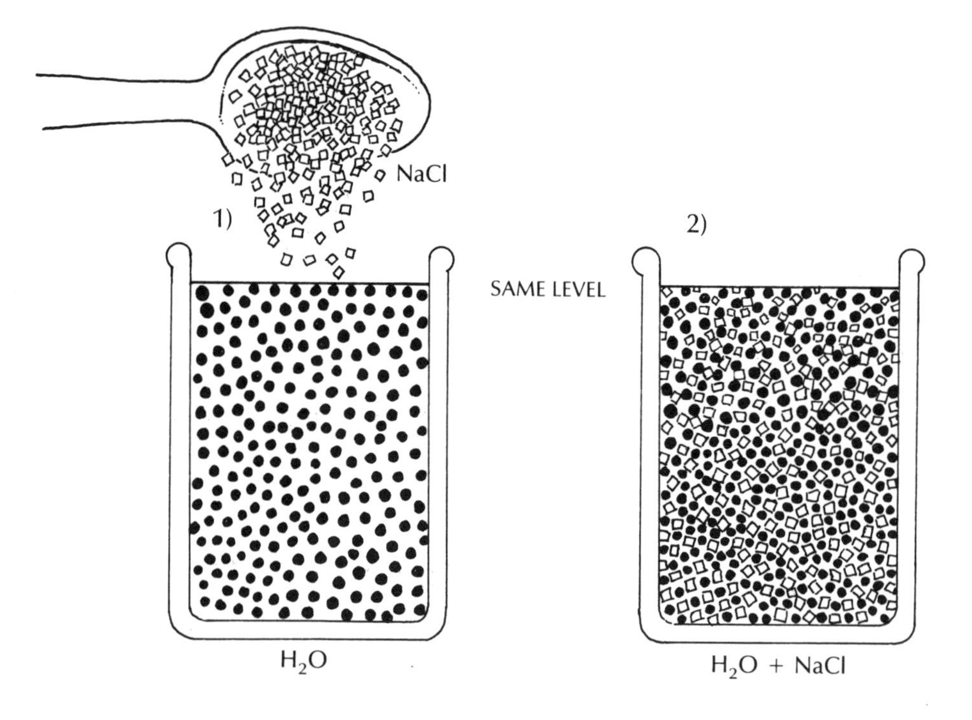


The molecules in liquids and gases, however, are not as tightly bound to one another. The molecules are farther apart and the cohesion is less strong. The molecules in liquid move about somewhat; the molecules in gases move freely. Other molecules can easily enter the spaces between the molecules of liquids or gases.

You can try this out. Put exactly 1 cup of water in a measuring cup. Be sure that the water level comes up right to the line marked on the cup. Now take a heaping teaspoon of salt and dissolve it in the water.

Common sense tells you that adding the salt to a full cup of water will raise the height of the water. Check it out. You'll see that the water did not go above the 1-cup line. The reason is that the salt molecules fit in between the loose water molecules. Therefore they did not take up any extra space.

Some substances can be either solid, liquid or gas, depending on the degree of heat. When you heat a substance, heat energy causes its molecules to move faster. Eventually, the molecules will move so fast that their cohesive forces cannot hold them together. As a result, the molecules separate. When you heat liquid water, for instance, you make the molecules shake about so much that the liquid becomes water vapor, a gas. On the other hand, cooling takes away energy. It increases the cohesive forces. Thus, when you cool liquid water it becomes ice, a solid.

The materials around you are changing all the time. There are two main kinds of changes—physical and chemical. One type of physical change is a change in the amount of space a material takes up. This is its volume. When you stretch a rubber band or blow up a balloon you are making a physical change in volume. When you take a handful of snow and pack it into a tight snowball you are also bringing about a change in volume.

Some physical changes occur when a material changes its state. You are familiar with water in its three different states. Ice is water as a solid. Liquid water is, of course, a liquid. And water vapor is a gas.

Here is an easy way to see how water changes its volume as it changes its state. Get a clean, empty container such as those used for margarine or cottage cheese. Fill it with water to the very top and carefully set it in the freezer part of your refrigerator. Look at it again after the water has frozen. You'll see that the solid ice sticks out above the top of the container. This shows that solid ice takes up more space than liquid water. (Water is the only common material that expands when it freezes. Everything else grows smaller in volume.)

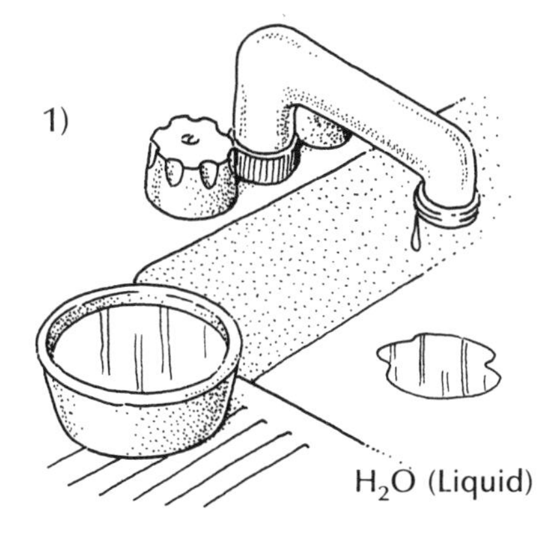


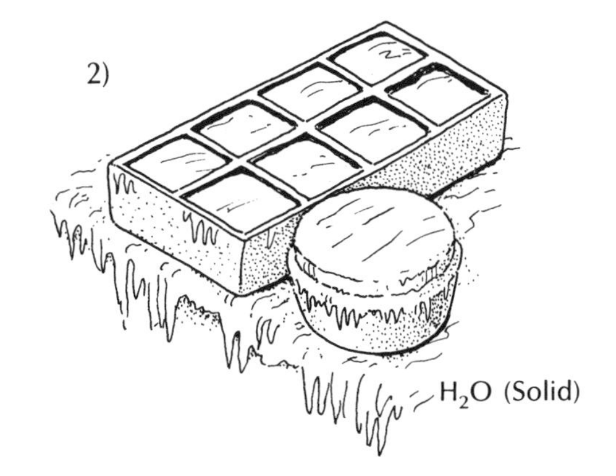

When you apply heat to a substance, the heat makes the nearby molecules shake and dart around. Each moving molecule bumps into other molecules. This makes them move about and spreads the heat throughout the material.

You can observe how moving molecules send heat through a piece of metal. You'll need a paper clip, a thumb tack, a match or lighter, a cork bottle-stopper and some candle wax. Straighten out the paper clip and squeeze a few small balls of wax onto the wire near one end. Set the center

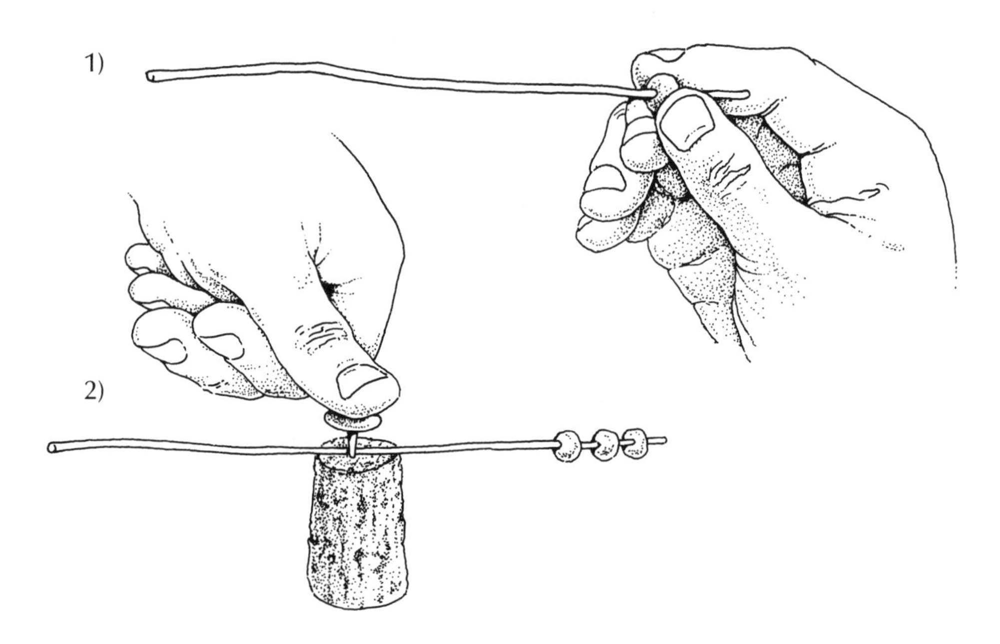


of the paper clip over the cork and use the thumb tack to hold it in place.

Now carefully light the match and hold it under the end of the paper clip opposite to the balls of wax. The heat energy makes the molecules in that part of the paper clip shake back and forth. The movement passes through the metal to the other end. When it gets there the heat melts the candle wax. The little balls of wax fall off.

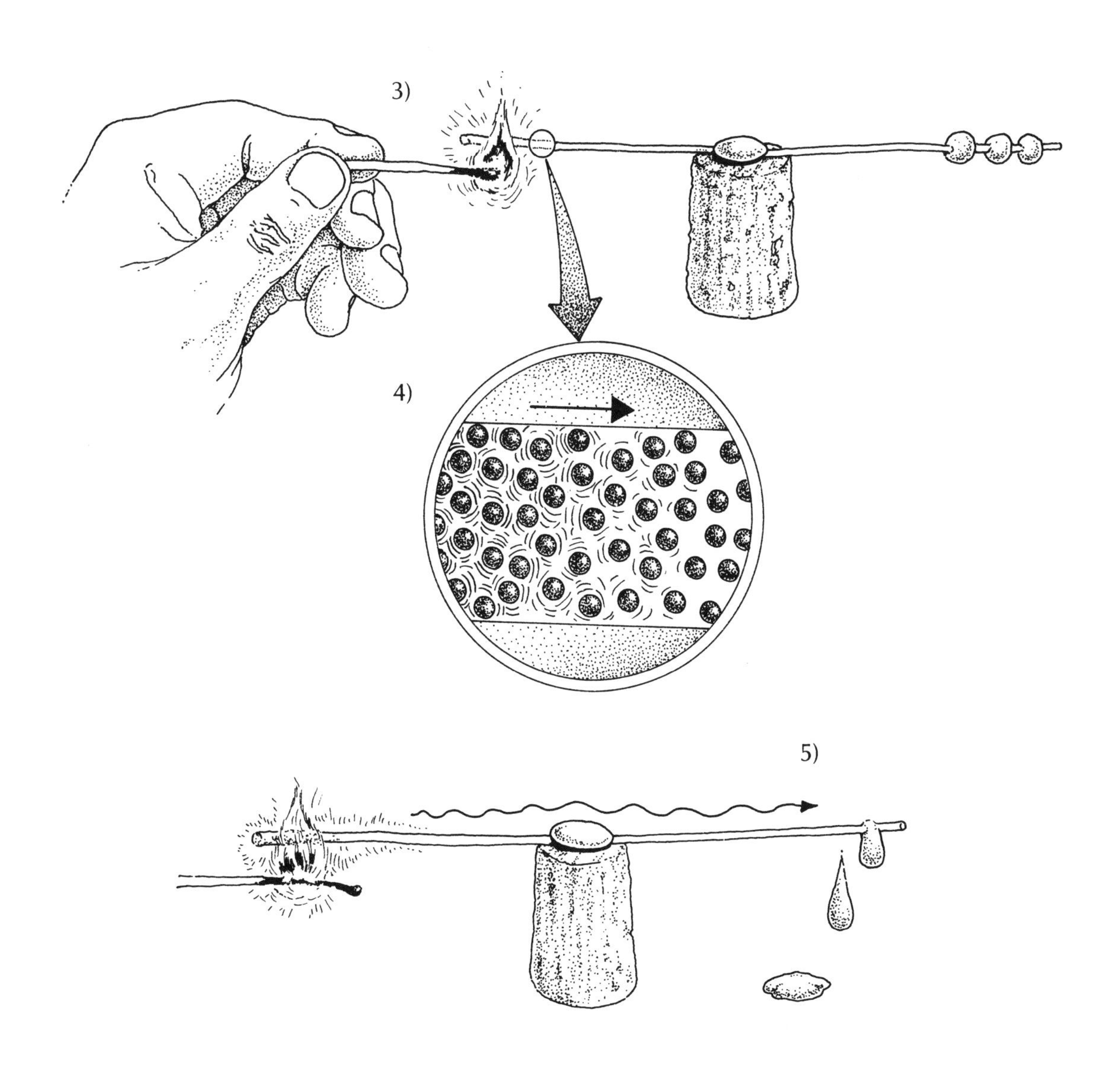
3)
4)
5)

Combining the molecules of two substances is another kind of physical change. Sometimes, though, the process can be reversed. You can prove this for yourself. Dissolve a tablespoon of salt in half a glass of water. The salt seems to disappear in the water.

Now pour the salt and water mixture into a pan and heat the mixture. The heat changes the liquid water into water vapor, which goes off into the air. When all the water is gone, you'll see some white crystals left in the pan. Taste some when they have cooled. You'll find that they are crystals of salt.

When a chemical change takes place, the molecules are actually changed into other molecules. They are different from the original ones. The atoms are arranged in a new way. Of course, the number and kind of atoms stay the same.

If you have some tarnished silverware you can cause a helpful chemical change in the silver molecules. You'll need water, an aluminum pan, baking soda and salt. Put 2 quarts of water into the pan. Add 2 teaspoons of baking soda and 2 teaspoons of salt. Place the pan on the heat and bring the mixture to a boil. Turn off the heat.

Carefully set each piece of silverware into the pan. Make sure that each one touches the pan and is covered by the water. Leave the silverware there for a few minutes. Then take out each piece, rinse in plain water and dry it with a soft cloth. Do you see how the tarnish is gone and the silver shines?

The tarnish is actually the compound silver sulfide ($Ag_2S$); two silver (Ag) atoms joined with a sulfur (S) atom. It is formed on the silver surface by different sulfur compounds in the air or in foods. When you placed the tarnished silverware in the pan, through a complicated series of chemical changes,the sulfur left the silver. The result was a chemical change from silver sulfide molecules to pure silver. And pure silver is always bright and shiny.

### *The Different Shapes of Molecules*

Many elements occur naturally in the earth in chemical compounds called ores, and have to be extracted from their ores. Some exist in natural compounds that can be used as they are.

Today, though, many of our most useful materials are compounds that have been created by scientists from natural elements. Compounds that are made by scientists are known as synthetic compounds. Plastics, for example, are synthetic compounds that are made from chemicals in petroleum and coal. Most of the dyes we use to change colors in materials are synthesized from chemicals found in coal tar. The chemicals for making synthetic rubber come from petroleum or natural gas. Various other synthetic compounds are prepared from chemicals in alcohol.

Most synthetic compounds contain the element carbon. Carbon forms far more compounds than any other element. The carbon atom has 2 electrons in its inner shell and 4 more in its outer shell. Since the outer shell can contain up to 8 electrons, the carbon atom can combine with as many as four other atoms. For that reason it can be used to create a fantastic variety of compounds.

Many thousands of different carbon compounds contain only the elements carbon and hydrogen. They are known as hydrocarbons. One of the simplest is methane, the main part of natural gas. In the methane molecule, the single electrons from four hydrogen atoms fill in the carbon atom's outer shell to the stable number of eight. The formula for methane is $CH_4$. The carbon and hydrogen atoms are linked in methane, and all the other carbon compounds, by covalent bonding.

Suppose you combine 2 carbon atoms with 6 hydrogen atoms. You get the next higher molecule in the hydrocarbon series, ethane ($C_2H_6$). This is followed by propane ($C_3H_8$). And so on and on, all of them linking very long chains of carbon molecules.

Ethylene ($C_2H_4$) belongs to another series of hydrocarbons. In this series there is a double bond between one pair of carbon atoms. A double bond means that each atom contributes two electrons to the link. That means that four electrons are involved in the bond between a pair of atoms. This link is stronger than a single bond.

In other groups of hydrocarbons, three electrons each from a pair of carbon atoms are joined, forming a triple bond. Acetylene ($C_2H_2$) is such a molecule. Another type of hydrocarbon, benzene ($C_6H_6$), is an example of a ring formation. Here the bonding forms the carbon and hydrogen atoms into a doughnutlike circle.

Crude petroleum consists of a mixture of many molecules. These molecules usually contain from six to several hundred carbon atoms. To make useful products from petroleum, scientists need to separate the various molecules from one another. This may be done by distillation. The crude petroleum is heated at the bottom of a tall still. As the vaporized petroleum rises through the still, the heavier molecules, such as lubricating oils, condense near the bottom. The lighter molecules, such as kerosene, condense higher in the tower. Gasoline, the lightest of all, condenses at the top.

In this way we are able to separate the petroleum into a whole series of

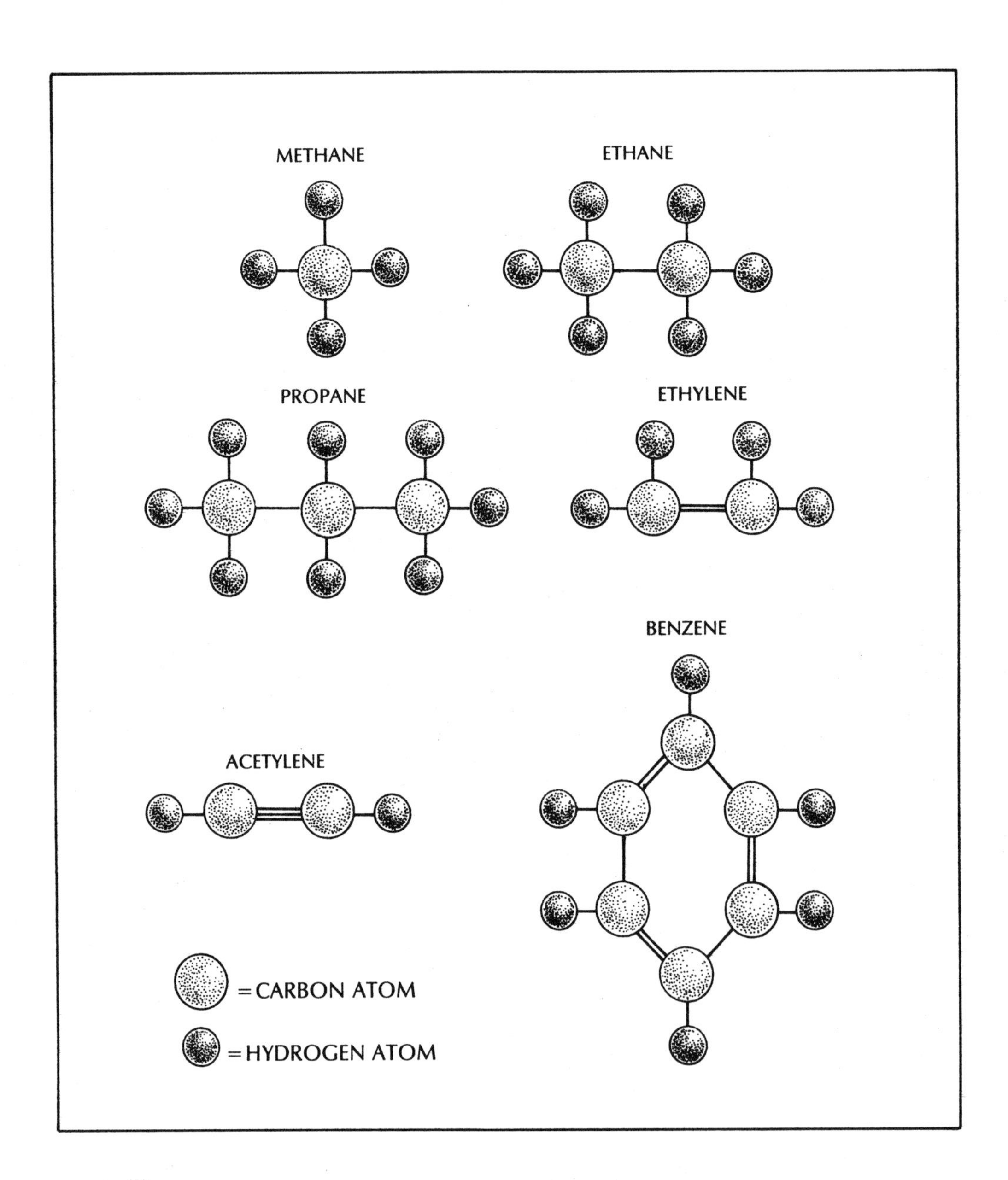
METHANE
ETHANE
PROPANE
ETHYLENE
BENZENE
ACETYLENE
=CARBON ATOM
=HYDROGEN ATOM

fractions. The process depends on how easy it is for the various molecules to boil away from the hot petroleum.

The lighter fractions, like kerosene and gasoline, are very useful as fuels. The heavier ones, however, are not very useful, except as greases and oils. For other purposes the atoms in these heavy molecules must be rearranged.

This is done by heating the large, heavy molecules under pressure in an oven. The process breaks them apart or cracks them into smaller pieces. The cracked pieces are put through a separation process in the still. The results are relatively pure fractions of the various cracked parts. From these fractions we can make many useful products, from fuels to plastics, from fibers to drugs.

The molecules made from petroleum can be quite large. Any such giant compound built up from molecules of the same kind is called a polymer. *Polymers* actually have different physical characteristics than do the individual molecules. Nylon, Dacron, Orlon, Teflon, polyethylene and polystyrene are some of the best known of the manmade polymers. There are also some natural polymers, such as rubber and cellulose (the main ingredient of plant cell walls).

Polymers exist because many molecules have their own electrical polarity. That is, one end of the molecule has a positive electrical charge and the other has a negative electrical charge. Therefore the positive end of one is attracted to the negative end of another. This binds them together, molecule to molecule. These molecules form into long chains.

Polymer scientists are able to create various types of giant molecules. With these molecules they can produce materials with many different characteristics. All of our plastics, synthetic fabrics and synthetic rubbers are polymers, made by scientists for specific purposes. They can be made stronger, more rigid, more elastic or more resistant to heat, stains, chemicals or whatever by the way the atoms and molecules are arranged.

Two groups of new synthetic polymers are becoming increasingly important. In one group, the scientists replaced the hydrogen in hydrocarbons with the element fluorine. The new compounds are known as fluorocarbons. The fluorocarbons are valuable because they are not affected by acids and other powerful chemicals. The fluorocarbons were widely used in industry until it was found that they destroy the ozone layer in the atmosphere. In 1977 their use was banned for some purposes, such as the propellant in aerosol products.

The other group is made up of materials called silicones. In these compounds, scientists put the element silicon in place of the carbon. Silicones are not affected by either high or low temperatures. Some have a freezing point as low as 75 degrees below zero F. Yet their boiling point and kindling temperature are above 500 degrees F. These properties make silicones useful in all kinds of weather. Oils, greases, varnishes and types of synthetic rubber are now made from silicones. They are also excellent insulators for wires in generators, transformers and motors.

Scientists today are using their knowledge of molecules to create exciting new materials that improve life for people all over the world. Many technological advances in medicine, engineering and daily living would not be possible without the remarkable molecules now in the service of humanity.

# 3. QUARKS

***What Quarks Are***

In the early years of this century, atoms seemed to consist of three basic particles—electrons, protons and neutrons. Scientists believed that each particle had a specific purpose in the atom. And they held that no other particles were necessary to explain atomic behavior.

But researchers continued to study the particles within atoms. There was always the question of whether these particles were made of even smaller, more basic building blocks.

Soon after the middle of the century, Murray Gell-Mann (b. 1929), a leading physicist at the California Institute of Technology, came up with one of the most startling theories of twentieth-century science. He proposed that the proton and neutron are each made up of three even tinier particles. He named these new particles quarks (pronounce it to rhyme with corks, not sparks).

Gell-Mann got the name *quark* from James Joyce's book *Finnegans Wake*. In the book, the phrase "three quarks" has two meanings. Sometimes it refers to Mr. Finnegan's three children. At other times it refers to Mr. Finnegan himself. In the same way, quarks can have two meanings. They can be thought of as separate, independent particles. Or they can be considered together as making up a larger particle, such as the proton or neutron.

Protons and neutrons are combinations of three different kinds of quarks. Gell-Mann gave them playful names. He called the three types, *up, down* and *sideways* quarks, even though they have nothing at all to do with directions. (Later the third one's name was changed to *strange* instead of sideways.)

From the time Gell-Mann first proposed the idea of quarks in the early 1960s, scientists have been trying to isolate a single quark. All efforts have failed. (In 1977 some physicists at Stanford University announced that they had found a single quark, but there has been no confirmation of their discovery.) But most scientists do not think that the failure to find one quark by itself disproves the existence of quarks.

The search for a single quark can be compared to trying to isolate one pole of a magnet. No matter how short you cut a bar magnet, it always has a north pole and a south pole. In the same way, it may be impossible to separate out an individual quark.

### *Flavors and Colors*

As scientists learned more about the three types of quarks, each one was found to have its own mass and electrical charge. The term *flavor* was used to describe the differences in these properties. That is to say, quarks have either up, down or strange flavor. These flavors are not at all related to the common flavors, such as vanilla or chocolate. They are only ways to identify the different kinds of quarks.

Recently scientists decided that quarks of the same flavor could not all be the same. They concluded that each flavor of quark can come in three different forms. These different forms are known as *colors*.

As with flavors, the quark colors have nothing to do with the colors we know. It is just a way of classifying quarks according to the way they join with other quarks. Different scientists use different sets of color names. Red, blue and green is perhaps the most popular set of quark colors.

With the idea of quark colors came another notion, that quarks are held together in the larger particles by something called the *color force*. You know that electrical charges attract or repel each other. So, too, do quark colors attract or repel each other. There is one difference, though. There are only two electrical charges, positive and negative. But there are three different quark colors, red, blue and green.

Scientists explain the working of the color force this way: There is another particle that actually carries the force. Since it glues the quarks together, this particle is called a *gluon*. In some ways the gluon acts like a particle. But since it has no mass, it also acts like pure energy. Either way, the gluon is the force that holds the quarks in a larger particle together.

The quark colors, we said, come from the different ways that quarks can combine. Scientists have discovered what they call the color rule. The rule states that every particle containing quarks is colorless. That means that in particles made up of three quarks there must be one quark of each color. That is to say, protons and neutrons contain a red quark, a blue quark and a green quark. They are colorless because the colors cancel each other out. Although the quark colors have nothing to do with the familiar colors, the ordinary colors can help to understand quark colors. Imagine blending equal amounts of red, blue and yellow paint in a dish. The color that results will actually be white, or colorless.

Research also shows that for every quark there is an identical quark with the opposite color (an anticolor) and an opposite electrical charge. These quarks are known as *antiquarks*.

### *The Family of Quarks*

Physicists learned a great deal about quarks between 1964 and 1974. Then in 1974 they found a new particle, the J-psi. The J-psi was important because its characteristics and behavior did not fit with any of the known quark flavors. Therefore scientists had to create another quark flavor. They gave the new flavor the name *charm*.

The quark "family," though, was still not complete. Three years later another new particle, upsilon, was found. None of the quark flavors seemed to apply here either. So still another quark was created. It was called *bottom* by some and *beauty* by others.

In the early 1980s, scientists decided that quark flavors came in pairs—up and down, strange and charm. So there had to be at least one more quark. It was named either *top,* to go with bottom, or *truth,* to go with beauty.

The entire family of quarks has one unusual feature in common. They all have fractional electrical charges. At one time, all the known particles had unit charges. The charge could be either positive or negative. But it was always a whole charge. Quarks, though, are different. The up, charm and truth quarks have a positive two-thirds charge. The down, strange and beauty quarks have a negative one-third charge.

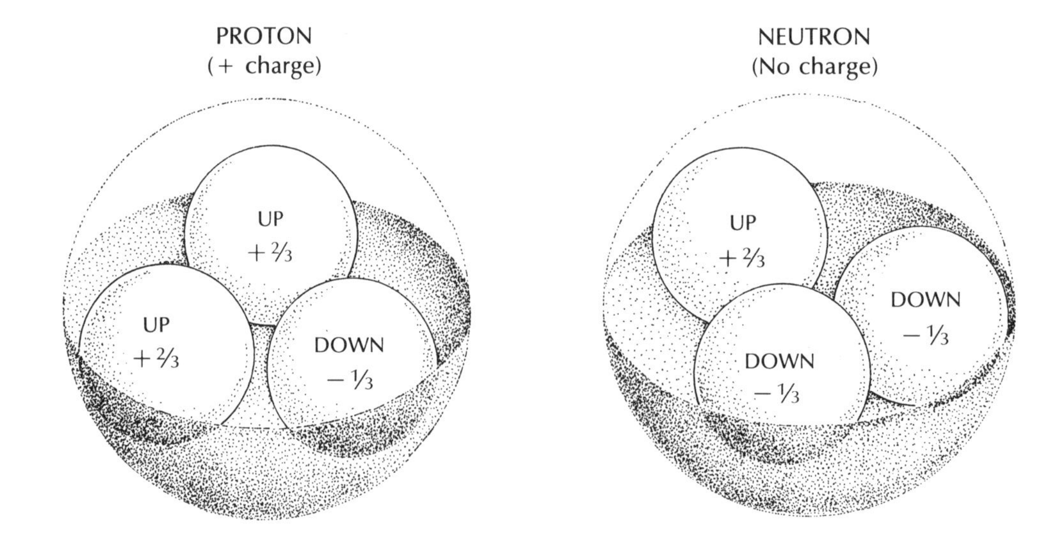

The proton, for example, has a single positive charge. It is made up of 3 quarks, 2 ups and a down. The 2 ups are $+\frac{2}{3}$ and $+\frac{2}{3}$, and the down is $-\frac{1}{3}$. This adds up to $\frac{3}{3}$, or 1, the proton's single positive charge.

The neutron has no electrical charge. It consists of 1 up and 2 down quarks. That is, $+\frac{2}{3} - \frac{1}{3} - \frac{1}{3}$, which equals zero. That is to say, the neutron has no charge.

Today the quark family consists of eighteen known members. There are the six basic flavors, up and down, strange and charm, truth and beauty, each of which comes in three colors, red, blue and green. In addition there are the equal and opposite antiquarks for every one of the flavors and colors.

**Quark Family**

| *Up Red* | *Up Blue* | *Up Green* |
| --- | --- | --- |
| *Down Red* | *Down Blue* | *Down Green* |
| *Strange Red* | *Strange Blue* | *Strange Green* |
| *Charm Red* | *Charm Blue* | *Charm Green* |
| *Truth Red* | *Truth Blue* | *Truth Green* |
| *Beauty Red* | *Beauty Blue* | *Beauty Green* |

### *The Particle Zoo*

Some researchers have been looking for ever smaller, more basic units of particles found within the atom. They have, as we said, already found eighteen different quarks.

Other scientists, meanwhile, follow a different course. They have been searching for additional particles within the atom. These experts have uncovered about two hundred "new" subatomic particles. With so many particles, it is easy to understand why the collection is sometimes called a particle zoo!

Scientists found most of the subatomic particles with immense modern machines called accelerators, or atom smashers. Accelerators use very powerful magnetic fields to send electrically charged atomic particles along tracks at speeds approaching the speed of light. These fast-moving particles are then used to bombard the nuclei of various atoms, smashing them to bits.

The results of the collisions give researchers important information,

*These two giant machines send particles through the accelerator at speeds near the speed of light.*

adding to their knowledge of the speeding particles and of the target particles and also disclosing the existence of completely new particles.

Using atom smashers to study atomic structure is very difficult. Some compare this approach to finding out how watches work by smashing two of them together, then picking over the pieces. One of the major problems is to be able to track the old particles and detect the new ones. Some of these particles disappear almost immediately, within billionths of a second!

Experts use two special devices to study the subatomic particles from the accelerator. One, the *spark detector,* is a mesh of fine wires that spark whenever struck by a particle. The other, the *bubble chamber,* is a huge structure filled with a heated liquid. When particles from the accelerator pass through the bubble chamber, they leave trails of tiny bubbles. The tracks in the bubble chamber are photographed continually. The particle trails are then studied to identify both the old and the new particles.

The spark detector and bubble chamber have largely replaced the *cloud*

*This photograph taken in a bubble chamber shows a proton and an antiproton destroying each other.*

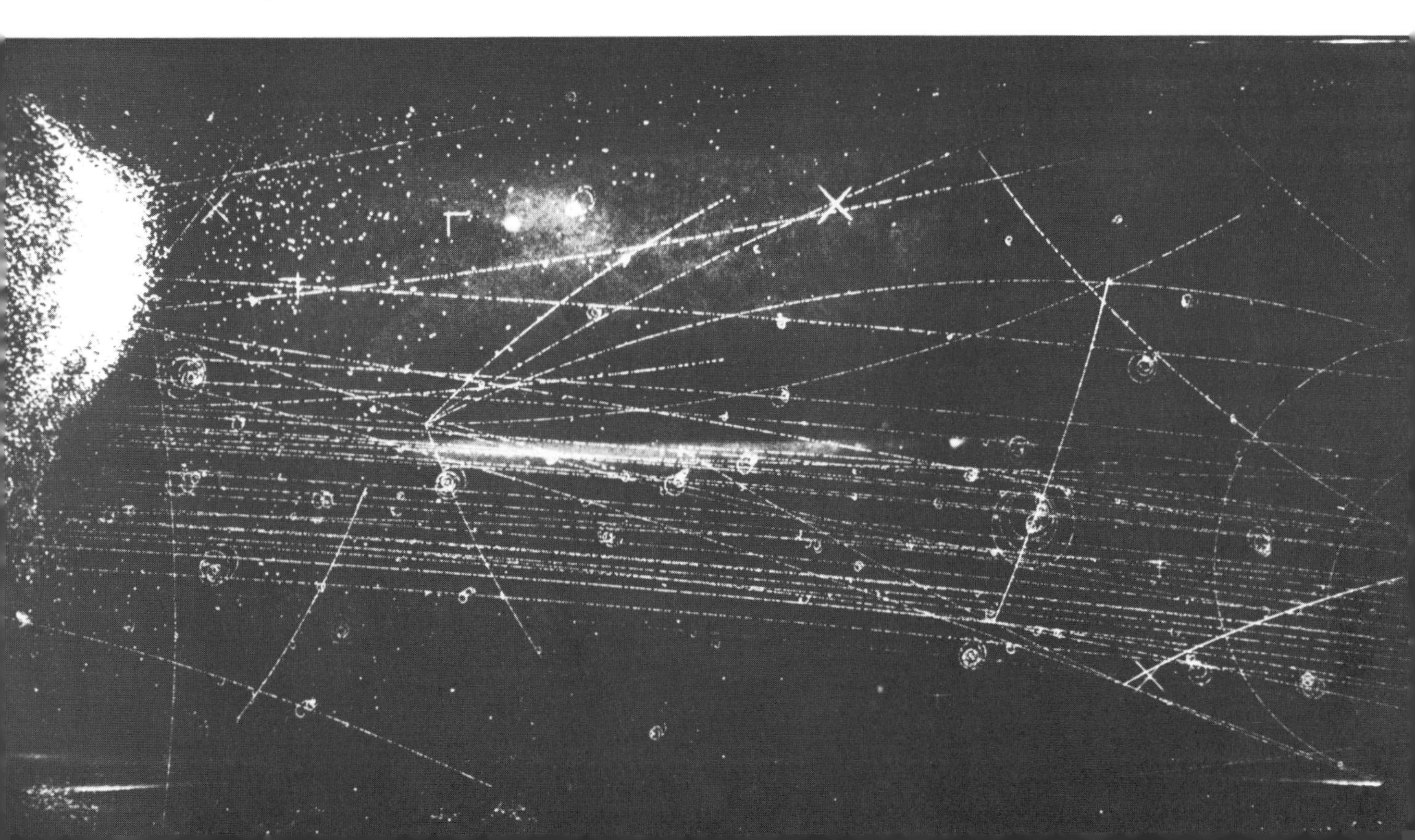

*This accelerator has a half-mile long track for the particles. Sometimes scientists ride bicycles to get around the track.*

*chamber*, which was used for many of the early experiments in nuclear physics. The cloud chamber works in much the same way as the newer device. The difference is that the particles leave vapor trails, not tracks of bubbles.

You can build your own cloud chamber. With it you'll be able to trace the paths of cosmic rays from outer space and follow the trails of particles from radioactive sources on earth. For supplies you'll need a large, wide-mouth jar with a screw-on metal cap, glue, some rubbing alcohol, a piece of dark black cloth, a piece of dry ice (look up "dry ice" in the Yellow Pages of your phone book for a source), a large, thin sponge or piece of felt and a source of bright, directed light, like a slide projector or powerful flashlight.

1. Find a jar with a tight cover, such as an empty, clean, dry peanut butter jar. The jar must be absolutely leakproof. One with a metal cover that has a rubber gasket is best.

2. Cut out a piece of the black cloth to fit inside the metal cover without covering the rubber gasket. Glue the cloth to the inside of the metal cover. Make sure the glue is completely dry before starting the actual experiment.

3. Cut out a piece of the sponge or felt to cover the inside of the bottom

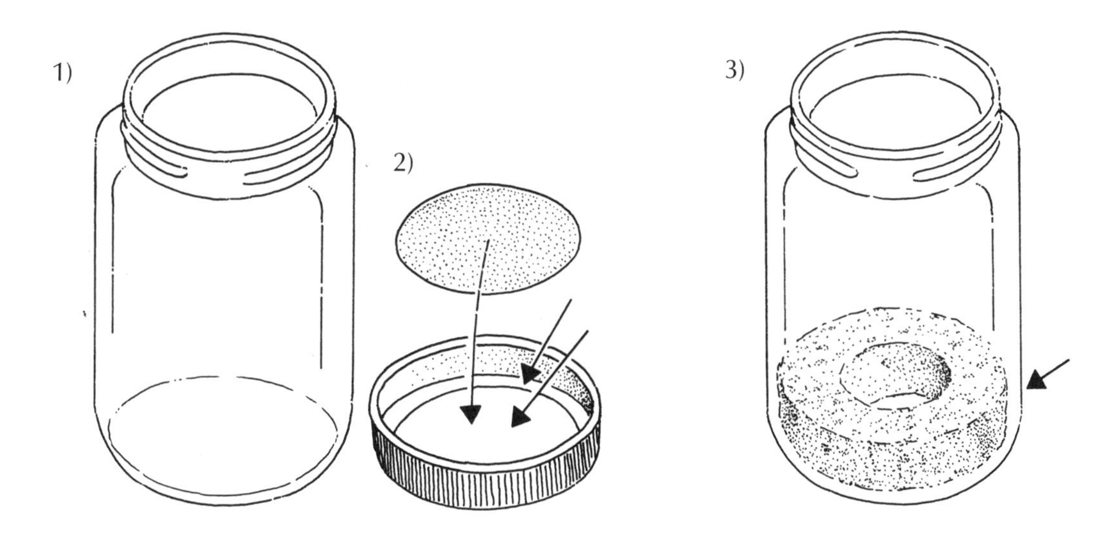

of the jar. Then make a doughnutlike hole in the middle. Glue the specially cut piece of sponge or felt in the bottom of the glass jar. Allow the glue to dry thoroughly.

4. Place one or two tablespoons of rubbing alcohol in the jar. Use enough to completely soak all of the sponge or felt.

5. Place the cover on the jar and twist it on very tightly.

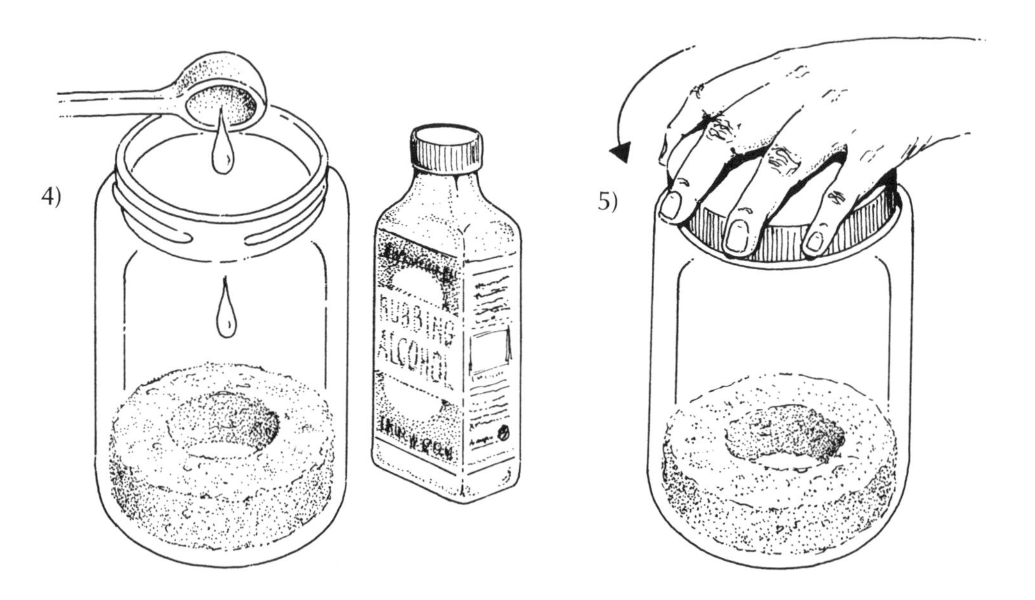


6. Turn the jar over so that the metal cap is on the bottom. Place it on the cake of dry ice.

7. Direct a bright, well-focused beam of light through the side of the jar toward the bottom (really the metal top) of the jar. A slide projector light works well. A flashlight can be used if the beam is bright enough and has a diameter of no more than about 1 inch.

8. Wait about 5 minutes. Observe the cloud or fog in the jar. Look either from the side opposite the light or through the doughnut hole in the bottom

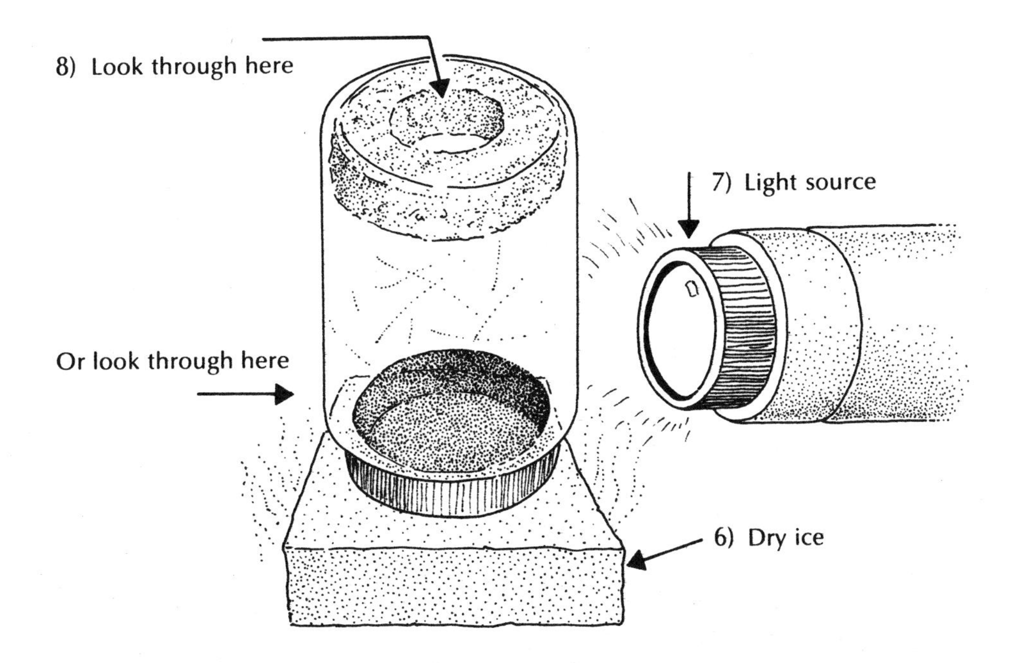


(now the top) of the jar. Watch especially carefully near the bottom. Be on the lookout for faint vapor trails that flash through the alcohol vapor and then quickly disappear. These paths are formed by particles, such as cosmic rays or radioactive beams coming from outside and passing through your cloud chamber.

If there is no fog in your chamber, either (1) the bottom is not cold enough, and you have to wait a bit longer, or (2) there is not enough alcohol, and you have to add some more to the sponge or felt.

If you see fog but no tracks, either (1) outside air is leaking in, and the cover must be made tighter, or (2) the inside of the jar is contaminated by the cloth, sponge or felt or by the glue, and you have to let the cloud chamber air for a day before using it.

The accelerators, as well as cloud and bubble chambers, have made possible the discovery of hundreds of particles over the last half century. Of all of them, the positron, neutrino, photon and meson are perhaps the most familiar.

### *Positrons*

The positron was discovered in 1932 by American physicist Carl Anderson (b. 1905). Anderson was then studying cosmic rays at the California Institute of Technology. In his experiments he observed the paths made by these rays from outer space as they passed through his cloud chamber.

To learn more about the particles that make up the cosmic rays, Anderson placed powerful electromagnets around the cloud chamber. The way the path bent told him whether the particle had a positive or negative charge. And the amount it bent gave him an idea of the particle's mass.

As he studied the various paths in the cloud chamber he noticed one very strange particle. It had the same mass as the electron, but it had a positive electrical charge instead of the electron's usual negative charge. He named the new particle *positron,* short for positive electron.

Scientists began studying the positron. They found it has a very short life. As soon as it meets an electron, which happens in a short fraction of a second, the particles destroy each other. Both the positron and the electron disappear.

Now scientists began to wonder if there were other particles like the positron. They expected them to be opposite in charge to the known particles and to destroy each other whenever they met. They called these particles *antiparticles.*

Eventually they found that there is an antiparticle for every particle. For those with electrical charges, like the electron or proton, the antiparticle has the opposite electrical charge. For those with no electrical charge, such as the neutron, the antiparticle has its spin or magnetic field turned in the opposite way. In all cases, though, when a particle meets its antiparticle, both particles disappear and there is a release of energy.

On the basis of their research, scientists believe there should be equal numbers of particles and antiparticles. Yet, if this is so, the particles and antiparticles would have destroyed each other long ago, and the universe would have disappeared in a burst of energy right after it was formed.

There seem to be two possible explanations for why antiparticles have not destroyed our universe. One is that the particles and the antiparticles moved apart right after the formation of the universe. Therefore, there may be some undiscovered part of our universe where there are vast galaxies made up mostly of antiparticles.

Another idea is that there was just a slight surplus of particles over antiparticles at the birth of our universe. In the billions of years since then, though, the particles slowly increased in number until today they vastly outnumber the antiparticles. At this time there is still no real proof for either of these theories.

### *Neutrinos*

In 1931, Austrian physicist Wolfgang Pauli (1900–1958) predicted the existence of a new particle, which he called the *neutrino*. Pauli believed the neutrino would explain the mysterious loss of energy that occurs in certain neutrons.

When a neutron is outside the nucleus of an atom, it goes through a change called decay. Each neutron is transformed into a proton and an electron. But in the decay of the neutron, scientists found that there was also a loss of energy. They asked the question, "What happened to the missing energy?"

Pauli tried to find an answer. He was a theoretical scientist; he never did experiments in a laboratory. He sometimes worked at a blackboard or at a desk where he pondered the unsolved problems and unanswered questions of physics. From time to time he hit on explanations and answers.

The solution for the neutron decay mystery, Pauli suggested, was that there was a particle that no one had yet detected. This particle, he theorized, carried away the lost energy. He called the mystery particle a neutrino, meaning "little neutron." Pauli figured out the following about the neutrino: It travels at close to the speed of light. It spins on its axis, just like the earth. It has no electrical charge. And finally, strangest of all, although it has no mass, it does have momentum, just like any moving object!

Pauli's description of the neutrino was very convincing. Yet no researcher could detect one. The reason? Neutrinos almost never react with other matter. A neutrino, on average, would have to pass through a solid block of lead 21 billion, billion miles thick before it interacted with another atom!

The existence of neutrinos was finally proved by two American physicists, Clyde Cowan (1920–1974) and Frederick Reines (b. 1918). In 1953 they set up huge detecting tanks next to a nuclear reactor in South Carolina to look for neutrinos. Their method was not to look for them directly, but to look for the neutrino's antiparticles, called antineutrinos. The presence of antineutrinos would indicate that neutrinos also exist. Using this method, they located the trail of an antineutrino. After four years of work, they confirmed Pauli's original theory.

### Photons

You know that negative and positive electrical charges are attracted to each other. This attraction is called the electromagnetic force. The electromagnetic force is carried by particles known as *photons*.

The electromagnetic force, working through the photons, keeps electrons in orbit around the nuclei of all atoms. As an example, think of the hydrogen atom. It has one proton in the nucleus and one electron spinning in orbit around the nucleus. The proton and electron have opposite electrical charges. The attraction of the electromagnetic force, or photon, keeps the atom intact.

Photons are similar to gluons. The gluons carry the color force between quarks. The photons carry the electromagnetic force between electrical charges. Both gluons and photons can be thought of either as particles or as energy.

In the hydrogen atom, the photons bounce back and forth between the proton and the electron. This keeps the two particles in place. The English-born, American physicist Banesch Hoffmann (b. 1906) compares a hydrogen atom to two tennis players (a proton and an electron) having a volley with a dozen balls (photons) flying in both directions.

The photons that carry the electromagnetic force are *virtual* photons. The term suggests that they are not real photons. All light is made up of photons. Virtual photons, though, cannot be seen as light. The only way virtual photons can become visible light is if their energy level is raised.

Imagine, then, a tennis volley in a court with a solid wall around it. No one on the outside can see what is happening. But if one of the players makes a very powerful stroke (provides more energy), the ball (the photon) goes up higher than the wall. It could then be seen by those on the outside (visible light).

### *Mesons*

The discovery of the photon as the carrier of the electromagnetic force led the Japanese scientist Hideki Yukawa (1907–1981) to wonder whether a similar particle carries the nuclear or strong force. This is the force that holds the protons and neutrons together in the atomic nucleus.

You know that according to the laws of electromagnetism, like charges repel each other. Yet in the nucleus the protons (all with the same positive charge) and the neutrons are held tightly together.

Scientists decided that there was a force able to overcome the electromagnetic force. Since it had to be very powerful it came to be called the strong force. They also realized that it could work only over short distances, like those inside the atom's nucleus. That is why it is also known as the nuclear force.

In 1935 Yukawa predicted that a particle, somewhere between the electron and the proton in mass, carried the strong force within the nucleus. Three years later two American scientists at the California Institute of Technology, Carl Anderson and Seth Neddermeyer (b. 1907), discovered a particle that fit Yukawa's description.

The particle's mass was more than 200 times that of the electron. This placed it about halfway between the electron and proton. So they called it a *mesotron* from the Greek word for middle. Later the name was shortened to *meson*.

Still later, scientists found two types of mesons. They added Greek letter prefixes to show the difference. One was called a *pi-meson,* which was shortened to *pion*. The other, *mu-meson,* became *muon*.

The pion is the particle that Yukawa believed was the carrier of the strong force. Pions have 270 times the mass of the electron. They may have positive or negative charges or be neutral. They have no spin. And they have an average lifespan of about one hundred-millionth of a second. The muon is more closely related to the electron, and is not now considered a meson.

### *Hadrons and Leptons*

By now hundreds of different particles have been found. They make up the "particle zoo." The scientists working with all these particles have set up a system to organize them into groups. It is the same idea zoo keepers use to classify their animals into types and classes according to their similarities and differences.

Particles that are affected by the strong or nuclear force are grouped together. These particles are all relatively large, and are all made up of quarks and antiquarks. They are called *hadrons,* from the Greek word for stout or strong.

The hadrons are further divided. The ones with the greatest mass are known as *baryons* (from the Greek word for heavy). The best known baryons are the protons and neutrons. The rest of the baryons have Greek letter

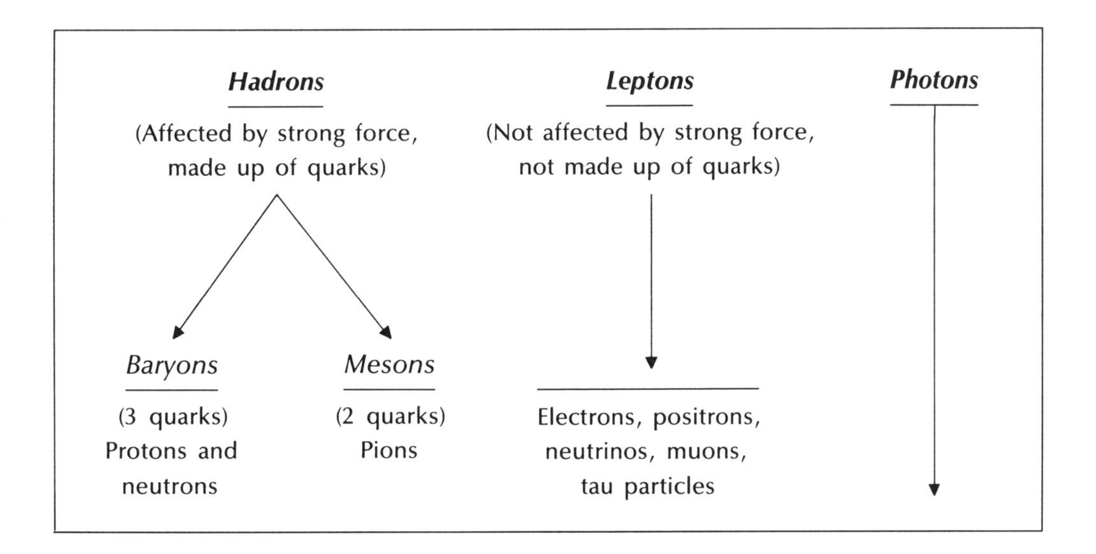


names, such as lambda, sigma and xi. Each baryon is believed to consist of three quarks.

The other group of hadrons are the *mesons*. The pion is the most familiar meson. The mesons are thought to contain only two quarks.

The second major group of particles is made up of those that are not affected by the strong force. These particles are tiny and are not made up of quarks. They are called *leptons* (the Greek root means small or tiny). The leptons are the smallest of the particles, with little or no mass. Among the leptons are electrons, positrons, neutrinos, muons and the recently discovered tau particles.

All the known particles and their antiparticles fit into one or the other of these groups. The only exception is the photon. It is unique and does not fall into any category.

Much recent research in physics has been aimed at finding a few basic building blocks of the universe. But the results have been far from simple.

Today there are eighteen known quarks and a growing list of two hundred particles.

No one knows what the future holds. Will the lists of quarks and other particles grow longer and more complicated? Or will a simple order be revealed? Will a new particle replace the quark as the basis of all matter in the universe?

These are just a few of the many fascinating questions that face modern scientists. But solving the mysteries of atoms, molecules and quarks will give us a better understanding of our truly incredible universe.

# GLOSSARY

***Adhesion*** The molecular force that makes one material stick to another.

***Alpha particle*** The nucleus of the helium atom, containing 2 protons and 2 neutrons.

***Antiparticle*** A particle with the same mass and the same characteristics as a standard particle, but with the opposite electrical charge. The antiproton is the same as the proton, except that the antiproton has a negative electrical charge. The antiparticle of the electron is the positron.

***Atom*** The smallest unit of a chemical element. Each atom consists of a small, dense central nucleus containing protons and neutrons and shells of circling electrons at a distance from the nucleus.

***Baryon*** Type of hadron. All baryons consist of three quarks. The most common baryons are protons and neutrons.

***Beauty*** One of the six types, or flavors, of quarks. Sometimes called bottom.

***Bottom*** One of the six types, or flavors, of quarks. Sometimes called beauty.

***Bubble chamber*** A device used to detect rapidly moving particles. The bubble chamber is filled with liquid, and when a particle passes through it leaves a track of bubbles, which can be photographed.

***Charm*** One of the six types, or flavors, of quarks.

***Cloud chamber*** An older device used to detect rapidly moving particles. The cloud chamber is filled with a gas vapor, and when a particle passes through it leaves a vapor trail, which can be photographed.

***Color*** Three different basic properties of all quarks. They are usually named red, blue and green, although they have nothing at all to do with the actual colors.

***Color force*** The force, carried by gluons, that holds the quarks together within a larger particle.

***Compound*** A substance that can be broken down into two or more elements.

***Cosmic rays*** High-energy particles that strike the earth from outer space.

***Covalent*** A form of electron bonding in which the atoms in a molecule share some of their electrons.

***Decay*** The change of an atom, nucleus or particle into two or more objects, whose total energy is less than that of the original object.

***Down*** One of the six types, or flavors, of quarks.

***Electromagnetic force*** One of the basic forces of nature, seen in the behavior of electrically charged particles. It is carried by the exchange of photons. The electrons and the nucleus are held together in atoms by the electromagnetic force.

***Electron*** The negatively charged particle of very small mass that normally circles the nucleus.

***Electrovalent*** A form of electron bonding in molecules in which one or more electrons pass from one atom to another.

***Element*** A simple substance made up of only one kind of atom.

***Flavor*** The qualities of a quark that determine which of the six types of quark it is.

***Gluon*** Particle that carries the color force between quarks.

***Hadron*** A particle made up of quarks that is subject to the strong force. Baryons and mesons are the two types of hadrons.

***Ion*** An atom with either more or fewer electrons than protons, giving the atom either a negative or a positive electrical charge.

***Ionic bonding*** Another name for electrovalent bonding. See also *electrovalent*.

***Isomer*** A compound with the same atoms as another, but with a different structure or arrangement of the atoms.

***Isotopes*** Two or more atoms of an element whose nuclei contain the same number of protons, but different numbers of neutrons. Isotopes are different in mass and nuclear properties, but have the same chemical properties.

***Lepton*** A tiny particle not made of quarks. Electrons, neutrinos, positrons, muons and tau particles are leptons.

***Mass*** The amount of matter in an object. It is related to weight, which is the pull of gravity on the object's mass.

***Meson*** A type of hadron that carries the strong force within the nucleus. Mesons are made of two quarks. The pion is a well-known meson.

***Molecule*** The smallest unit of matter made up of two or more atoms.

***Muon*** A lepton. In some ways, the muon is really a heavier electron.

***Neutrino*** A particle without mass and without electrical charge that carries away the energy lost in neutron decay. A neutrino is a lepton.

***Neutron*** One of the two particles found in the nucleus of all atoms except one isotope of hydrogen. A neutron has slightly more mass than the proton, but has no electrical charge.

***Nuclear force*** See *strong force*.

***Nucleus*** The positively charged central core of all atoms, where almost all of the mass of the atom is concentrated. It contains protons and neutrons.

***Particles*** In general, tiny bits or pieces of matter. In science, often used as short for elementary or fundamental particles, referring to the parts found within the atom.

***Photon*** A particle without mass that travels at the speed of light. Photons are exchanged when electromagnetic force is applied. All light is made up of photons.

***Pion*** The lightest meson. It is responsible for the strong force in the nucleus.

***Polymer*** A compound formed of a long chain of simpler molecules.

***Positron*** The antiparticle of the electron. It has same mass as the electron but with a positive electrical charge equal to the negative charge of the electron.

***Proton*** One of the two particles found in the nuclei of all atoms. A proton has a positive electrical charge, and is approximately equal in mass to the neutron.

***Quantum theory*** The accepted theory that radiation is not emitted continuously, but in small, separate units. Each unit is called a quantum.

***Quark*** The particle believed to be the basic building block of several atomic particles.

***Radioactivity*** The property of some elements, such as radium, of emitting particles or radiation or both.

***Spark detector*** A device used to follow the paths of particles coming from a particle accelerator.

***Spin*** The idea that atomic particles spin as the earth does on its axis.

***Strange*** One of the six types, or flavors, of quarks.

***Strong force*** One of the basic forces in nature. The strong force holds the protons and neutrons together in the nucleus of the atom. The strong force is much stronger than the electromagnetic force, which would cause the positively charged protons in the nucleus to repel one another and fly apart.

***Subatomic*** A reference to any particle found within an atom.

***Top*** One of the six types, or flavors, of quarks. Also called truth.

***Truth*** One of the six types, or flavors, of quarks. Also called top.

***Up*** One of the six types, or flavors, of quarks.

# INDEX